Sir Francis Galton, FRS

The Legacy of His Ideas

Proceedings of the twenty-eighth annual symposium of the Galton Institute, London, 1991

Edited by

Milo Keynes
Galton Institute, London

D1363372

M in association with
THE GALTON INSTITUTE

First published 1993 by
THE MACMILLAN PRESS LTD
Houndmills, Basingstoke, Hampshire RG21 2XS
and London
Companies and representatives
throughout the world

ISBN 0-333-54695-4

A catalogue record for this book is available
from the British Library.

Printed in Great Britain by
Antony Rowe Ltd
Chippenham, Wiltshire

STUDIES IN BIOLOGY, ECONOMY AND SOCIETY

General Editor: Robert Chester, Department of Social Policy and Professional Studies, University of Hull

Published titles

Milo Keynes, David A. Coleman and Nicholas H. Dimsdale (*editors*)
THE POLITICAL ECONOMY OF HEALTH AND WELFARE

Peter Diggory, Malcolm Potts and Sue Teper (*editors*)
NATURAL HUMAN FERTILITY

Milo Keynes and G. Ainsworth Harrison (*editors*)
EVOLUTIONARY STUDIES: A Centenary Celebration of the Life of Julian Huxley

David Robinson, Alan Maynard and Robert Chester (*editors*)
CONTROLLING LEGAL ADDICTIONS

D. F. Roberts and Robert Chester (*editors*)
MOLECULAR GENETICS IN MEDICINE: Advances, Applications and Ethical Implications

A. H. Bittles and D. F. Roberts (*editors*)
MINORITY POPULATIONS: Genetics, Demography and Health

Milo Keynes (*editor*)
SIR FRANCIS GALTON, FRS: The Legacy of His Ideas

Series Standing Order

If you would like to receive future titles in this series as they are published, you can make use of our standing order facility. To place a standing order please contact your bookseller or, in case of difficulty, write to us at the address below with your name and address and the name of the series. Please state with which title you wish to begin your standing order.
(If you live outside the United Kingdom we may not have the rights for your area, in which case we will forward your order to the publisher concerned.)

Customer Services Department, Macmillan Distribution Ltd, Houndmills, Basingstoke, Hampshire, RG21 2XS, England.

STUDIES IN BIOLOGY, ECONOMY AND SOCIETY

General Editor: Robert Chester, Department of Social Policy and Professional Studies, University of Hull

The Galton Institute is concerned with the interdisciplinary study of the biological, genetic, economic and cultural factors relating to human reproduction, development and health in the broadest sense. The Institute has a wide range of interests which includes the description and measurement of human qualities, human heredity, the influence of environment and the causes of disease, genetic counselling, the family unit, marriage guidance, birth control, differential fertility, infecundity, artificial insemination, termination of pregnancy, population problems and migration. As a registered charity, the Institute does not propagate particular political views, but it does seek to foster respect for human variety and to encourage circumstances in which the fullest achievement of individual human potential can be realised. More generally, the Institute seeks to advance understanding of biosocial matters by enabling biologists, clinicians, demographers, sociologists and other professionals to work together in a mutually productive manner.

The Galton Institute was formed in 1989 as the successor body to the Eugenics Society, which in turn derived from the Eugenics Education Society founded in 1907. Membership of the Institute is international and consists of Fellows and Members. Fellows are those who contribute by their work and writings to the advancement of knowledge in the biosocial sciences. Members are drawn from a wide area of biosocial interests. Amongst its activities, the Institute supports original research via its Stopes Research Fund, sponsors the annual Darwin lecture in Human Biology, co-sponsors the biennial Caradog Jones Lecture, and is associated with the Biosocial Society in publishing the *Journal of Biosocial Science*. Each year the Institute mounts a two-day symposium in which a topic of current importance is explored from differing standpoints, and during which the Galton Lecture is delivered by a distinguished guest. The proceedings of the symposia since 1985 constitute the successive volumes of the series *Studies in Biology, Economy and Society*. The balance between disciplines varies with the nature of the topic, but each volume contains authoritative contributions from diverse biological and social sciences and an editorial introduction.

Information about the Institute, its aims, activities and publications may be obtained from: The General Secretary, The Galton Institute, 19 Northfields Prospect, Northfields, London, SW18 1PE.

Contents

Notes on the Contributors

W. H. G. Armytage is Emeritus Professor of Education at the University of Sheffield.

Michael Banton is Professor of Sociology at the University of Bristol and Past President of the Royal Anthropological Institute.

W. F. Bynum is Head of the Academic Unit at the Wellcome Institute for the History of Medicine, London and Reader in History of Medicine, University College London.

A. W. F. Edwards is Reader in Mathematical Biology at the University of Cambridge and Fellow of Gonville and Caius College, Cambridge.

J. H. Edwards, FRS, is Professor of Genetics at the University of Oxford and Consultant in Medical Genetics, Oxford Regional Health Authority.

H. J. Eysenck is Emeritus Professor of Psychology at the University of London and former Director of the Department of Psychology, Institute of Psychiatry, University of London.

Gertrud Hauser is Associate Professor at the Histologisch-Embryolisches Institut der Universität, Wien.

J. S. Jones is Professor of Genetics and Head of the Galton Laboratory and Department of Genetics and Biometry, University College London.

Milo Keynes is a member of the Galton Institute, London.

C. G. Nicholas Mascie-Taylor is Head of the Department of Biological Anthropology at the University of Cambridge and Fellow of Churchill College, Cambridge.

John Maynard Smith, FRS is Emeritus Professor of Biology at the University of Sussex.

Dorothy Middleton is Honorary Vice-President of the Royal Geographical Society.

J. M. Tanner is Emeritus Professor of Child Health and Growth at the University of London.

Sir Crispin Tickell is Warden of Green College, Oxford, and President of the Royal Geographical Society.

Introductory Note

Sir Francis Galton, who lived from 1822 to 1911, was a polymath with a 'universal scientific curiosity', though his scientific success was achieved despite receiving no adequate scientific education and only obtaining a pass degree in mathematics at Cambridge. He travelled in Africa and the Middle East, and on his return used his strong mechanical bent to improve scientific instruments employed in geographical work. He then changed to meteorology, devising weather maps and discovering the anticyclone.

He was a grandson of Erasmus Darwin, and the appearance of *On the Origin of Species* by his cousin Charles altered his life when it demolished 'a multitude of dogmatic barriers by a single stroke'. He immediately saw the implications for the history of man, and began to wrestle with the idea of evolution and its application to mankind. From 1865 his whole effort was to try to discover laws of inheritance, at first using rough methods for analysing them, inventing crude ranking statistics such as percentiles. He rejected Quetelet's idea that all variation in human physical characteristics was error about a type, that all deviation from the average was error. He saw that variation was the fuel for natural selection, that without it there was no evolution, and tried to account for it. His first attempt was, like that of Mendel, by using peas, the yield of which he classified, creating what was probably the first bivariate distribution. From this he constructed the first regression line, calling it 'reversion', in 1877, and then moved on to collect data that was more refined, by opening an anthropometrical laboratory for the study of human variation. The discovery of the coefficient of correlation followed. He was concerned to distinguish the effects of heredity, or nature, from those of the environment, or nurture.

Galton was the founder of human genetics and of the science of statistics in Britain, and the founder of eugenics, endowing a professorship in this subject at the Galton Laboratory, University College London. The *Eugenics Education Society* was founded in 1907, with Galton its honorary president from the next year until he died in 1911. This became the *Eugenics Society* in 1926, and there was a further change in name to that of the *Galton Institute* in 1989. This book is the publication of the proceedings of the Institute's annual two-day symposium on *Sir Francis Galton FRS – The Legacy of his Ideas* held in 1991, eighty years after Galton's death.

MILO KEYNES

1 Sir Francis Galton – A Man with a Universal Scientific Curiosity

Milo Keynes

In a letter of 1871, Charles Darwin (1809–1882) wrote:

> I have been speculating last night what makes a man a discoverer of undiscovered things; and a most perplexing problem it is. Many men who are very clever – much cleverer than the discoverers – never originate anything. As far as I can conjecture the art consists in habitually searching for the causes and meaning of everything which occurs. This implies sharp observation and requires as much knowledge as possible of the subject investigated.

Both Darwin and his cousin Francis Galton had this compulsive curiosity. John Maynard Keynes (1883–1946), in some remarks before his Galton Lecture of 1937, considered, indeed, that

> there was no-one who has possessed in purer essence than he [Galton] the spirit of universal scientific curiosity . . . he had a rare and delicious fancy, and his job was, so to speak, to put all of us on enquiry – into a thousand fertile corners (M. Keynes, 1984).

Francis Galton, the founder of human genetics and of the science of statistics in Britain, as well as the founder of eugenics, was a true polymath, when the widely diversified nature of his other, often pioneering, scientific interests are included. The only connecting thread, however, between his widespread contributions to science appears to have been his wayward personality. Eliot Slater (1904–1983), in his Galton Lecture for 1960, thought 'he was a scientific dilettante at his super-excellent best'. But, it must be added, Keynes had already remarked:

> It was not the business of his particular kind of brain to push anything far. His original genius was superior to his intellect, but

1

his intellect was always just sufficient to keep him just on the right side of eccentricity.

Several contributors to this book have commented that he did not always show the necessary drive to reach the full, deserved conclusion from his important discoveries.

Galton never received an adequate scientific education. For two years before going to Cambridge University to read mathematics, he worked as a medical student at the Birmingham General Hospital as well as studying anatomy and physiology at King's College, London. But he only learned enough mathematics at Cambridge to attain a Poll (pass), not an Honours, degree. He had no apprenticeship, no mentor, and never knew the discipline of work in a research laboratory. He worked on his own throughout his scientific life, and as his fancy took him so brilliantly.

'It has always been my unwholesome way of work', he wrote in *Memories of my Life* (1908), 'to brood much at irregular times'. He had the ability and good fortune, however, to brood on previously almost unexplored fields, owing little to the researches, published or communicated, of other men. As Slater (1960) commented, it has been rightly noted that discoveries of fundamental importance are often made by amateurs, since in a new and untouched field there can be no experts, though of course the amateur who makes discoveries in a new territory is usually an expert in some more established field. But Galton was an expert in nothing, and Karl Pearson (1857–1936), his biographer, could only write that at the age of thirty-two he knew more mathematics and physics than most biologists, more biology than most mathematicians, and more pathology and physiology than either.

Besides his 'ideas', which led him into so many strange paths and gave him his enthusiasm for the British Association for the Advancement of Science, of which he was honorary secretary from 1863 to 1867, Galton had a strong mechanical bent, producing over the years a long succession of mechanical inventions, such as a printing electric telegraph, an instrument for sun-signalling, stereoscopic maps, spectacles for divers, the Galton whistle for testing the power of hearing for high notes, and a fingerprint examiner. At the Galton Laboratory at University College London there are a number of ingenious contrivances, 'Galton's toys', the purpose of many of which is unknown. Sir Francis Darwin (1848–1925), Charles Darwin's third son and biographer, gave the first Galton Lecture in 1914. In this he remarked that his father had a special affection for what he called his 'fool's experiments', but that Galton had omitted to record his own such experiments, many of which, Francis Darwin felt, would have been in delightful lines of work.

Galton became a Fellow of the Royal Society in 1860 for his work on scientific instruments used in geographical research, such as helping to devise means for standardising sextants and regulating chronometers, particularly after he became a member of the Kew Observatory where the main interest was in magnetic observations. It was the friendships and connections he made during his scientific career that led to his subsequent scientific evolution. He developed an immense respect for his fellow-scientists – the reverence of the amateur for the professional. He wrote:

> A collection of living magnates in various branches of intellectual achievement is always a feast to my eyes; being as they generally are such massive, vigorous, capable-looking animals.

But of his own status he was markedly modest, with an extremely low self-esteem. Pearson noted that Galton regarded himself as one whose faculties only gave him rank in the extreme tail of a frequency distribution.

Few of Galton's books are available in modern reprints and his numerous papers are lost in nineteenth-century periodicals. In 1908, at the age of eighty-six, he published *Memories of my Life*. Karl Pearson, the first Galton Professor at the University of London, produced the massive, 2000 page, three volume work, *The Life, Letters and Labours of Francis Galton* in 1914, 1924, and 1930. There is a second biography, *Francis Galton – The life and work of a Victorian Genius* by D. W. Forrest, that was published in 1974 and from which a bibliography of Galton's published work has been taken for this book. In 1982 a memoir of Sir Francis Galton by Dorothy Middleton was published by the Eugenics Society to celebrate the seventy-fifth anniversary of its founding. The subject of Chapter 2 of this book is 'The Historical Galton' by W. F. Bynum.

THE EARLY YEARS

Francis Galton, known to his family and friends as Frank, was a schoolboy of fifteen when Queen Victoria came to the throne, having been born in Birmingham on 16 February 1822, and had outlived her by ten years when he died on 17 January 1911. His grandfather, Samuel John Galton FRS (1753–1832), made a large fortune from the manufacture of muskets. Despite being disowned by the Quakers for this occupation, he continued to attend Quaker meetings. According to Galton (1908), he was 'a scientific and statistical man of business' who in his spare time was devoted to scientific self-education. An account of his work on colour vision appeared in *Monthly Magazine* in 1799, antedating the trichromatic theory of Thomas Young MD, FRS (1773–1829) by two years. He was a friend of Josiah Wedgwood FRS (1730–1795) and a member of the Lunar Society, of which Erasmus Darwin MD, FRS (1731–1802) was a founder member as well as Josiah Wedgwood's physician.

In 1804 Samuel John started the Galton Bank in Birmingham. He was partnered by his eldest surviving son, Samuel Tertius (1783–1844), of his family of six sons and four daughters. Samuel Tertius, Frank's father, besides being a wealthy banker had 'a scientific bent', as well as, according to Frank, being 'eminently statistical by disposition' – among other things he wrote a small book on currency. He married Violetta Darwin (1783–1874) in 1807, and the birth of five daughters was followed by the birth of two sons, Darwin (1814–1903) and Erasmus (1815–1909), and a daughter (1816–1817) before Frank was born six years later. At the death of two of his daughters in 1817, Samuel Tertius finally left the Quaker sect and was baptised into the Church of England. He retired from the bank when he was forty-eight and died when he was sixty.

The physician, poet and philosopher, Erasmus Darwin, was first married, in 1757, to Mary Howard (1740–1770), by whom he had three surviving sons. After she died he fathered two illegitimate daughters before marrying in 1781 a widow, Elizabeth Collier (1747–1832), whose first husband, Edward Pole (1718–1780), had been one of his patients. His second family consisted of four surviving sons and two daughters, of whom Violetta was the elder. Robert Waring Darwin MD, FRS (1766–1848) was the third son of Erasmus's first family. He married Susannah Wedgwood (1765–1817), Josiah Wedgwood's eldest daughter, in 1796 and they had a family of four daughters and two sons, of which Charles Robert Darwin was the second son. Frank Galton was thus Charles Darwin's half first-cousin, the younger by thirteen years, with Erasmus Darwin the grandfather of both.

Frank, the baby of the family, had four sisters who were between eleven and fourteen years older than he. It was the third, Adèle

(1810–1883), who particularly cared for him in childhood and seems to have begun his education, so that he could read a little book, *Cobwebs to Catch Flies*, by the time he was two and a half and a few months later could sign his name. He was undoubtedly precocious – he could tell the time and knew his pence and multiplication table by the age of four. Frank's Quaker background might have limited the choice of schools for him, but that does not account for his being sent at the age of eight to a wretched establishment at Boulogne, where his 'heart rebelled' for two years at the beatings and cheatings, the doleful walks and the lack of any learning. He then did better at a private school at Kenilworth, near to his new home, which had changed from Birmingham to Leamington on his father's retirement in 1831, where he was at least left to his own devices.

With his father's Quaker aversion to public schools, when he was fourteen he was sent to the indifferent Free School (now King Edward's School), Birmingham, which he disliked:

> The character of the education was altogether uncongenial to my temperament. I learnt nothing, and chafed at my limitations. I had craved for what was denied, namely, an abundance of good English reading, well-taught mathematics, and solid science. Grammar and the dry rudiments of Latin and Greek were abhorrent to me, for there seemed so little sense in them (*Memories*, 1908).

In this he suffered like his cousin Charles Darwin, who groaned over the classics of Shrewsbury School and forgot what he had learned, even some of the Greek letters, by the time he was nineteen.

At sixteen, in 1838, Frank was allowed to start a medical training at Birmingham General Hospital: the practice of medicine was a tradition in his mother's Darwin family. By this time nearly all his letters were addressed to his father, with whom he had a warm supportive relationship, and very few to his mother. In fact it appears his eldest sister Elizabeth (1808–1906) had taken over his mother's role: in his *Memories* of 1908 virtually all Francis Galton says of his mother is that 'she was very methodical in her papers and accounts, and a most affectionate mother to myself'.

At the hospital he was at first confined to the dispensary, where he began to sample the different drugs in alphabetical order to learn their properties. However he only got to the end of the letter C before being stopped by the diarrhoea caused by two drops of croton oil, which put to an end his thirst for first-hand knowledge. He had to prepare tinctures, extracts and decoctions and learned to make pills by hand – it was this latter facility that must have inspired, when later in life he saw a pill-making machine at work, the characteristic calculation that if the government possessed forty-five of them it could supply each inhabitant of the British Isles with one free pill a day. At the hospital he accompanied the surgeons in their rounds of the wards, watched operations and enjoyed bandaging and treating fractures. It was before the days of chloroform, and observing that an injured drayman, who had been brought in dead drunk with both of his thighs crushed and mangled, underwent amputations without any sign of feeling pain, he wondered whether patients might not, with advantage, routinely be made dead drunk before operations. At the end of the year he went to Germany and Austria to visit hospitals with two recently qualified doctors, one of whom was William Bowman (1816–1892), later a very eminent ophthalmic surgeon.

The next year (1839) he went to King's College, London and began to work in the dissecting room – where Bowman, who was working on the cellular structure of the human body, was then attached – and to study physiology. After the exams in April 1840, when he was '2nd Prizeman in Anatomy and Chemistry' but second in physiology, he gained the prize in forensic medicine in July. He became a frequent visitor to Gower Street to see Charles Darwin, recently married to Emma Wedgwood (1808–1896), youngest child of Josiah II (1769–1843) and granddaughter of Josiah I. As had Bowman, who told him that 'every high mathematical MD that he knows got on well', Darwin, who must have talked about his travels, and who had just published his *Journal of Researches* made during the voyage of *The Beagle*, recommended him to go to Cambridge the following October,

> to read Mathematics like a house on fire. . . . He said very truly that the faculty of observation rather than that of abstract reasoning tends to constitute a good Physician. The higher parts of Mathematics which are exceedingly interwoven with Chemical and Medical Phenomena (Electricity, Light, Heat etc, etc.) all exist and exist only on experience and observation (letter to Samuel Tertius, 6 December 1839).

That summer before Cambridge, Frank Galton made the first of his foreign tours when he moved on from the planned stay in Germany – in Giessen with the chemist Justus von Liebig (1803–1873) – and sailed down the Danube to Vienna, on through Budapest, Istanbul and the Black Sea to Syria, and then west to Athens, Venice and home.

In October 1840 Samuel Tertius, who had been at Trinity College forty years earlier but without taking a degree, went with Frank on the stagecoach to Cambridge. At that time undergraduates could only at first read for the Mathematical Tripos, after which they could then take the Classical Tripos, a subject in which, as already noted, Frank had no interest. In his first term at Trinity College he was ill with rheumatism and his studies were interrupted, so that he only got a third class in the college examinations in May. In his second year things were better, and especially enjoyable for him was the metaphysics of mathematics. But he then began to find it difficult to keep up in the fourteen different mathematical subjects, and complained of headaches and having what he called a 'sprained brain' due to overwork.

At the beginning of the third year he broke down, as was to be recurrent over the years, with symptoms suggestive of an anxiety state – headaches, dizziness, inability to concentrate, palpitations and obsessive thoughts – and had to go home for a period (letter to Samuel Tertius, 28 November 1842, and *Memories*, 1908). He gave up all effort to work for an honours degree and spent the time in social and literary activities, so his 'time was by no means wasted'. However he still attended the lectures necessary for a medical qualification. But with only a poll degree he was, he wrote in 1908, so pleased to be elected, in 1902, an Honorary Fellow of Trinity College (where his brother-in-law, H. Montagu Butler (1833–1918), had been Master since 1886) that he presented to the college the copy of his portrait by C. W. Furse (1868–1904) that is reproduced on page 21 of this book.

He returned to Cambridge for the fourth year, taking the poll degree examination for his non-honours BA degree in the January of his eleventh term. He stayed in Cambridge until June 1844 to continue his medical studies, which he did in a somewhat desultory way; after leaving Cambridge he attended some lectures at St. George's Hospital in London. By March 1844, however, his father had become ill and particularly needed Frank's medical care in the weeks before he died in October. It seems that his father, having escaped from banking and living the life of a scholar, had been pushing Frank towards a profession with which he was becoming steadily disillusioned on account of the humbug that would be necessary to be a successful general practitioner, or so he thought. On his father's death, he inherited sufficient wealth to be independent of any profession, and it was with relief that he ceased his medical studies.

TRAVEL AND WEATHER

In October 1845 Frank Galton travelled – this time in style – to Egypt and the Sudan, up the Nile to Khartoum, with a cousin and another Cambridge friend. On the way home they split company at Alexandria, and Galton arrived home in November 1846 after a visit to the Lebanon and Syria. Subsequently he shot and hunted, enjoyed long spells in London, and lived as an English country gentleman (Figure 1.1). In 1849 he was elected to the Royal Geographical Society, and began to make plans for an expedition to Africa, preferably to the south where Livingstone had recently crossed the Kalahari desert. He set sail for the Cape in April 1850 and, unable to go as planned to the unexplored region south of the Zambezi, took ship for South West Africa, landing in Walvis Bay and marching inland from there into Damaraland, now Namibia. Galton's African journey is the main subject of Dorothy Middleton's contribution to this book (Chapter 3).

It was on this African trip that Galton made an early contribution to the science of animal behaviour. While most of the cattle in a herd would stay close together with their horned heads turned outwards so that they could defend themselves against attack, a few grazed away from the herd, putting themselves at risk. When several pairs were hitched to a waggon, a lead pair of ordinary cattle tended to get frightened and snarl up the train, but the few cattle that grazed away from the herd made good lead oxen. Galton noted that in this world there were comparatively few men who had the courage and drive to 'graze away from the herd', and that these were sometimes badly mauled by 'human lions' (Galton, 1908).

Galton arrived back in England two years to the day from leaving it and was received with acclaim by the Royal Geographical Society, to whom he had sent an account of his expedition two months before (Galton, 1852). In fact he was lionised, 'which is exceedingly wearisome to the lion after the first excitement and novelty of the process have worn away' (Galton, 1908). In 1853 he wrote a longer account of his travels, and on 24 July received a congratulatory letter from Charles Darwin, who had lost touch with him, in which he wrote:

> I live at a village called Down near Farnborough in Kent, and employ myself in Zoology; but the objects of my study are very

Figure 1.1   Watercolour sketch of Francis Galton, aged 25, in 1847

Source: Wellcome Institute Library, London.

small fry, and to a man accustomed to rhinoceroses and lions, would appear infinitely insignificant (Darwin, 1853).

He became concerned with helping the Royal Geographical Society to advise intending travellers, and was awarded its founder's gold medal in 1854 (RGS, 1854). His famous *The Art of Travel*, which had an eighth edition in 1893, first appeared in 1855. He became a member of the council of the society, and its honorary secretary from 1857 to 1863, when he was forced to resign after a quarrel with the executive secretary, Norton Shaw. It was this that was probably instrumental in his being immediately offered the secretaryship of the British Association.

He first met his future wife, Louisa Butler (1822–1897), at the very beginning of 1853 and married her in August. She was the daughter of the Dean of Peterborough, Dr. George Butler (1774–1853), and sister of Montagu Butler, later Master of Trinity College. When Frank went to meet the dean for the first time to get engaged, he arrived only to find that the dean had died earlier that day, thus causing a delay in the rush into marriage. The marriage to Louisa, which was to be childless, was not a romantic one, but more of one into an intellectual family. In his *Memories* (p. 158) he wrote:

> I protest against the opinions of those sentimental people who think that marriage concerns only the two principals; it has in reality the wider effect of an alliance between each of them and a new family.

After an eight-month honeymoon and wintering in Italy, the Galtons returned to London to live in Victoria Street, SW1 until buying a house at Rutland Gate, South Kensington in 1857.

It was when he was in Africa that Galton particularly began to think about weather, and soon after his return he became a member of the managing committee of the Kew Observatory, later becoming its chairman from 1889 until it became fully merged in the National Physical Laboratory in 1901. But he did not move on to the consideration of meteorology until 1860, when he became aware of the work of the Meteorological Office, set up in 1854 by Vice Admiral Robert Fitzroy FRS (1805–1865), the former Captain of *The Beagle*, who conceived the Fitzroy barometer and initiated a system of storm-warnings and the first weather forecasts. After Fitzroy's suicide the Meteorological Office was reorganised into a Meteorological Committee, of which Galton was a member until his resignation in 1901.

Before this he had collected weather data from all over Western Europe and evolved the thematic mapping of weather on charts. His discovery of the anticyclone – the term 'cyclone' was originated in 1848 by Henry Piddington (1797–1858) – was first described in a paper read before the Royal Society in 1862 (Galton, 1863). This was followed by the publication in 1863 of *Meteorographica*, which contained 'Methods of mapping the weather, illustrated by upwards of 600 printed and lithographed diagrams . . . during the month of December 1861'. His weather maps appeared in *The Times* from April 1875. '*Meteorographica* and weather' is the subject of Sir Crispin Tickell's chapter in this book (Chapter 4).

ANTHROPOLOGY

The majority of the supporters who in 1837 formed the Aborigines Protection Society in London belonged to the Society of Friends, but soon there was a divergence of aim amongst its members. The majority wished to continue giving active political support and guarding the interests of native peoples living in the British colonies and dependencies, and were greatly concerned with the emancipation of slaves. The minority, who preferred to make a scientific study and understand the ways of life of the native populations, broke away in 1843 to found the Ethnological Society, which Galton joined in 1862 at the same time as the Darwinists, T. H. Huxley (1825–1895), John Lubbock, later Lord Avebury (1834–1913), and the surgeon George Busk (1807–1886), all of whom belonged to the evolutionary movement and were interested in investigating systematically the problems of man's origin and nature. It was a time when the slave question in the United States was a highly emotive issue and the secretary of the society, James Hunt (1833–1869), was striving to rouse his fellow-ethnologists to practical activity. In 1863 Hunt and his followers broke away to found the Anthropological Society, leaving Lubbock to become president and Galton secretary of the Ethnological Society. When Hunt died in 1869, negotiations were opened up between the two rival societies, and in 1871 they united to form the Anthropological Institute, again under the presidency of Lubbock. Galton himself was President from 1885 to 1889.

As an anthropologist Galton was predominantly concerned with man, developing an interest in what now might be called 'human biology'. But his first paper (Galton, 1883) to the Ethnological

Society in 1863 was concerned with the domestication of wild animals, in which he recognised that an animal's success in this direction, including the cat, depended upon whether the inborn and inherited mental qualities of the species of animal concerned fitted them for living as members of a herd or group. He had become a psychologist as well as an anthropologist. His next two relevant papers were published in 1865 in *Macmillan's Magazine* and are discussed in the next section. According to Sir Arthur Keith in his Galton Lecture of 1920, he had decided that they were not suitable for his fellow-ethnologists.

With the increasing interest in quantitative variation in man that followed the publication of *On the Origin of Species*, Galton became involved with anthropometry, but his first paper on anthropological statistics from schools, which recorded the measurements of the boys of Marlborough College, only appeared in 1874. In 1875 the Anthropometric Committee of the British Association was set up, with Galton first as secretary and later as chairman. The committee made its final report in 1883 after obtaining measurements from 53 000 people.

In 1884 he set up an anthropometric laboratory – an event that Skanes (1984) has called the beginning of experimental applied psychology – which he maintained and financed until 1891. He made appeals and offered prizes to obtain the records of 'family faculties', as he thought there was a 'dearth of information about the Transmission of Qualities among all the members of a family during two, three, or more generations', the statistics relating to 783 brothers and the records of thirty-five pairs of twins. He brought photography into his anthropological work not merely to record data, but as a means of obtaining 'generic' images by the combination of photographs of individuals of a family or of a group. He used composite photography not only for eliciting resemblances, but also for the analysis of the differences or individual characteristics of the individuals of a group. He drew up and distributed registers in which family data might be recorded. His work on anthropometry, which was what led to his being invited to become president of the Anthropological Institute in 1885, is discussed by J. M. Tanner in Chapter 8 'Galton on Human Growth and Form'.

Galton was not the first to open a laboratory for the measurement of human subjects: Alphonse Bertillon (1853–1914) had started such a venture for criminal investigation in Paris – the Department of Judicial Identity – two years earlier in 1882. But it was only in 1888 that Galton began experimenting with inked plates in taking fingerprints, and it was not until 1894 that he persuaded Bertillon to add fingerprinting to his system. In 1880 Henry Faulds, a medical mission-ary in Japan, had written to Charles Darwin and then to *Nature* that he had found racial differences in the pattern of fingerprints. He had defined three types of patterns and was convinced of their potential usefulness in the identification of criminals. Sir William Herschel, in reply to Faulds's letter in *Nature*, added his observations on the value of the use of fingerprints in the jails of Bengal for over twenty years. But Galton, in a lecture to the Royal Institute in 1888, at first overlooked this earlier work, and then later sought to credit the origination of the use of fingerprints for purposes of identification to Herschel, though it was Faulds who had done this and who had also suggested the basis of the classification system – in terms of the three types of pattern – adopted by Galton. Galton systematically collected fingerprints from the Anthropometric Laboratory, after finding out how to make them last more permanently, and established their uniqueness and importance in biological and criminological investiga-tion as well as how to classify them. His last major book, *Finger Prints*, came out in 1892, fifty-two years after his prize in forensic medicine as a medical student. Gertrud Hauser writes on 'Galton and the Study of Fingerprints' in Chapter 10 of this book.

HUMAN GENETICS AND THE SCIENCE OF STATISTICS

As is emphasised by W. F. Bynum in Chapter 2 of this book, it was when Galton read Darwin's *Origin of Species* when it appeared in 1859 – his annotated copy is in the strongroom at University College, London – that his life was altered.

> It made a marked epoch in my own mental development, as it did that of human thought generally. Its effect was to demolish a multitude of dogmatic barriers by a single stroke, and to arouse a spirit of rebellion against all ancient authorities whose positive and unauthenticated statements were contradicted by modern science (Galton, 1908).

He immediately saw the implications for the history of man, and began to consider mankind's future in the light of the theory of evolution. He later wrote:

> I was encouraged by the new views to pursue many inquiries which had long interested me, and which clustered round the central

topics of Heredity and the possible improvement of the Human Race (Galton, 1908).

His religious belief did not survive, and he describes how reading *Origin of Species* drove away 'the constraint of my old superstition as if it had been a nightmare' (Galton, 1908).

His first work on heredity appeared in 1865, after many months of hard work, under the title 'Hereditary talent and character' in two preliminary papers in *Macmillan's Magazine*. The work contained statistical proof of the inheritance of intellectual and moral qualities, though using statistical methods he was later to criticise, and told of the harm done by the Roman Catholic Church in recruiting as priests (and therefore to make them live as celibates) the intellectual, the literary and the sensitive (Galton, 1908). In the two papers, according to Francis Darwin (1914), Galton 'insists that the whole spiritual nature of man is heritable, so that in his opinion there are no traces of that new element "specially fashioned in Heaven" which is commonly believed to be given to a baby at its birth' – that human attribute commonly known as original sin. Galton discusses, too, the development of social virtues by natural selection, concluding that 'the development of our nature under Darwin's law of natural selection has not yet overtaken the development of our religious civilization', that is, he is considering the conflict between individual desires with tribal instincts. Later he came to differ from this viewpoint of 1865, believing that he was then 'too much disposed to think of marriage under some regulation, and not enough of the effects of self-interest and of social and religious sentiment' (Galton, 1908). (See Figure 1.2)

In 1866 Frank Galton had another breakdown which lasted on and off, with several long periods spent abroad, for two years. His next publication was *Hereditary Genius* (1869), later wishing he had used the word 'talent' rather than 'genius'. The thesis of the book is that 'genius', or exceptionally high ability, is genetically, rather than environmentally, determined. He developed a method for classifying ability and, noting that people differ in their abilities, claimed that the differences were innate. In the book he wrote:

> I have no patience with the hypothesis . . . that babies are born pretty much alike, and that the sole agencies in creating differencies between boy and boy, and man and man, are steady application and moral effort. It is in the most unqualified manner that I object to pretensions of natural equality. The experiences of

Figure 1.2 Francis Galton, aged 42, in 1864

Source: Wellcome Institute Library, London.

the nursery, the school, the university, and of professional careers, are a chain of proofs to the contrary (Galton, 1869).

The book was badly received by non-scientific readers, mainly because in it Galton criticised the Church and the clergy: his anti-clericalism reflected his loss of the religious faith of his Quaker upbringing. Scientists however liked it, though Charles Darwin, in a letter to Galton on 3 December 1869, made the reservation that he still maintained the *eminently* important difference 'that men did not differ much in intellect, only in zeal and hard work'. In the first edition (1871) of *The Descent of Man*, called a kind of triumphant family romance by Desmond and Moore (1991), Darwin asserted that man might, by selection, do something for the moral and physical qualities of the race.

In his Galton Lecture for 1983, H. J. Eysenck (1984) pointed out that Galton believed that intelligence was a scientifically meaningful concept underlying all cognitive behaviour, such as learning and problem solving, and that besides being largely heritable as well as socially important, it should be a measureable quantity, of use for instance when we want to select people in any way for the possession of a given trait or ability. The demonstration of heritability is also clearly dependent on having something that can be measured. Galton made several suggestions as to the best way of carrying out such measurement, including his view that reaction times would be a good measure. But beyond indicating the desirability of using physiological and relatively elementary measures he did not, as did Alfred Binet (1857–1911) after him, proceed to create usable scales of measurement. The measure of intelligence using the Binet test, a test that can be powerfully influenced by education, socioeconomic status and cultural factors, has led to never-ending arguments about the relative influence of nature and nurture. Such arguments have bedevilled, for instance, the scientific investigation of sex and class differences in intelligence, which the use of less specialised tests based on Galton's suggestions might have avoided (Eysenck, 1979). In Chapter 5 of this book H. J. Eysenck writes on 'Intelligence and Hereditary Genius'.

A short, somewhat provocative article in 1872, 'Statistical inquiries into the efficacy of prayer', which caused offence in clerical circles, was followed in 1874 by a book, *English Men of Science: their Nature and Nurture*, based on about a hundred of the replies to a lengthy questionaire of seven quarto pages sent to 180 selected Fellows of the Royal Society. Charles Darwin's answer contained the following:

Special talents, none, except for business, as evinced by keeping accounts, being regular in correspondence, and investing money very well; very methodical in my habits. Steadiness; great curiosity about facts, and their meaning; some love of the new and marvellous. . . . I suppose I have shown originality in science, as I have made discoveries with regard to common objects.

The next year, continuing his efforts to distinguish the effects of heredity from those of the environment, he brought out a paper (Galton, 1875) on the use of like-sex twins reared either together, or reared apart and therefore differently, to separate 'that portion of a man's nature which is due to heredity, from all the rest'. It is a misconception that Galton proposed, at this start of the study of human behavioural genetics, the comparison between identical and fraternal twins. 'Galton and the Use of Twin Studies' is the subject of C. G. N. Mascie-Taylor's contribution to this book (Chapter 9).

Galton's main effort from 1865 was to try to discover laws of inheritance. He used rough measurements to devise rough methods for analysing them, inventing crude ranking statistics such as percentiles. But, most importantly in his statistical thought, he rejected the idea of L. A. J. Quetelet (1796–1874) that all variation in human physical characteristics was error about a type, and accepted 'that the laws of Heredity were solely concerned with deviations expressed in statistical units' (Galton, 1908).

He saw that without variation there was no evolution and, realising that deviation from the average was not an error, tried to account for it. Encouraged by Charles Darwin, and with the backing of the botanist Joseph Hooker (1817–1911), he began to breed sweet peas, using them for the three reasons that they were hardy and prolific, had little tendency to cross fertilise, and all the seeds in the pods were roughly the same size. He classified the seeds according to size, gave them to seven friends, including Darwin, to plant, and described the yield according to both size of product and size of parent, to create what was probably the first bivariate distribution, from which he later constructed the first regression line, though calling it 'reversion' (Galton, 1877, 1885).

J. H. Edwards discusses the sweet pea work in his Galton Lecture printed here (Chapter 6), and points out that Francis Galton might have preceded Gregor Johann Mendel (1822–1884) in the scientific world's knowledge of his genetic law by twenty-three years had he been aided by a mathematician who was more of a biologist, or had he himself known more mathematics than he had acquired for his poll degree. Mendel performed his work on the edible pea from 1856–63, and presented it in two lectures in Brünn (Brno) early in 1865 (Mendel, 1865, with English translation, 1901). He had thought his results were inconclusive and needed more research, and only reluctantly accepted the offer of publication after reexamination of his calculations, calculations that R. A. Fisher (1890–1962) later subjected to statistical analysis and found to be too close to theoretical prediction to have occurred by chance alone (Fisher, 1936). The work lay dormant until 1900 when, within a two-month period, Hugo de Vries of Amsterdam, Carl Correns of Tübingen and Erich von Tschermak of Vienna independently arrived at the same conclusions as Mendel, and each discovered his paper in their preparations for publication. The wider recognition of 'Mendel's Law' was finally achieved on the appearance in 1909 of *Mendel's Principles of Heredity* by William Bateson (1861–1926). It was Bateson who in 1905 gave the name 'genetics' to the study of heredity and its variation.

Galton had realised that the inborn qualities transmitted by one generation to another proceed not from person to person, but from seed to seed. In his chapter on 'Galton and Evolutionary Theory' (Chapter 11), John Maynard Smith discusses Galton's experiments on Darwin's theory of pangenesis, his ideas on discontinuity in evolution – that evolution does not necessarily proceed in small steps – and his 'flirtation with Mendelian ratios'.

Inquiries into Human Faculty was published in 1883, and in it Galton reprinted a variety of papers, including those on what he called his 'psychometric experiments' and other psychological work, with a religious section, the history of twins and a chapter on 'nurture and nature'. The general object was:

> to take note of the varied hereditary faculties of different men, and of the great differencies in different families and races, to learn how far history may have shown practicability of supplanting inefficient human stock by better strains.

The book was concerned with the variety of human nature, particularly that of the comparative outward physical characteristics of human beings, and contained sections on composite portraiture and anthropometry and the need for more statistical enquiries. As mentioned, he considered mental and physiological factors, and felt strongly that the mental and physical ones were interrelated.

Galton's study of human variation continued to be statistical, and he made a further important discovery after his regression line. This was the 'coefficient of correlation', what he called an 'index of correlation', which was subsequently extended by Karl Pearson to variables not governed by the Gaussian law of distribution (Galton, 1888a). *Natural Inheritance* (1889), Galton's final expression of his ideas on the statistics of heredity, followed the next year, and of course was written before the discovery of Mendel's paper of 1865. In it Galton is concerned with three main questions: how do the characteristics of parents relate to those same characteristics in the offspring, what is the relative contribution of each ancestor to the nature of the offspring, and how is it possible to measure the nearness of kinship? (Forrest, 1974.)

Eliot Slater, in his Galton Lecture, thought that,

> Though the calculus of correlation has become the most powerful tool available to the psychometric psychologist, and is used as a basic instrument over a great variety of scientific fields, in conceptual importance it seems to me to come second to Galton's study of qualitative variation.
>
> When one is concerned with human attributes, the change of attitude involved in considering people as showing a characteristic to a greater or lesser degree, instead of merely having or not having it, is revolutionary. . . . It used to be universal . . . to think of people as being either "neurotic" or "not neurotic". When one thinks in terms of more and less, "neurosis" loses its unfortunate pathological implications and comes to be seen as a mode of normal variation (Slater, 1960).

Galton showed originality and made important contributions in the early years of the study of human genetics and statistical theory. In this book these are the subjects of the Galton Lecture for 1991 by J. H. Edwards (Chapter 6) as well as of Chapter 7 by A. W. F. Edwards.

EUGENICS

In *Memories of my Life* (1908) the last chapter, 'Race Improvement', defends Francis Galton's views on 'eugenics', about which he had first written – without using that name – in the two articles in *Macmillan's*

Magazine in 1865 while preparing *Hereditary Genius* (1869). He coined the word 'eugenics' and first used it in *Inquiries into Human Faculty* (1883), where he noted the word derived from

> the Greek *eugenes*, namely, good in stock, hereditarily endowed with noble qualities. This . . . [is] equally applicable to men, brutes, and plants. We greatly want a brief word to express the science of improving stock, which is by no means confined to questions of judicious mating, but which, especially in the case of man, takes cognisance of all influences that tend in however remote a degree to give to the more suitable races or strains of blood a better chance of prevailing speedily over the less suitable than they otherwise would have had.

Galton added that in the inquiries he made on the antecedents of men of science,

> no points came out more strongly than that the leaders of scientific thought were generally gifted with remarkable energy, and that they had inherited the gift of it from their parents and grandparents. . . . Energy is an attribute of the higher races, being favoured beyond all other qualities by natural selection . . . well washed and combed domestic pets grow dull; they miss the stimulus of fleas. . . . In any scheme of eugenics, energy . . . is eminently transmissible by descent (Galton, 1883).

It was not until 1901 that he returned to writing on eugenics, a subject that became the main preoccupation of his last ten years. His *Essays in Eugenics* appeared in 1909, the year he was knighted. In this he wrote that

> the aim of Eugenics is to represent each class or sect by its best specimens; that done, to leave them to work out their common civilisation in their own way. The list . . . when picking out the best qualities . . . would include health, energy, ability, manliness and courteous disposition (Galton, 1909).

Francis Darwin (1914) subsequently noted that the list might have served as a self-description.

Galton was not pleased when George Bernard Shaw commented that protagonists of eugenics would 'select their wives and husbands

Figure 1.3 Oil portrait of Sir Francis Galton, aged 81, in 1903

Source: Copy by F. W. Carter of painting by C. W. Furse presented to
Trinity College, Cambridge in 1905 (by permission of the Master
and Fellows).

far less carefully than they select their cashiers and cooks', and that
polygamy should be in order, as 'it seems a natural loss to limit the
husband's progenitive capacity to the breeding capacity of a single
woman' (Shaw, 1905). Shaw was too extreme and deliberately provocative, while Galton was preaching caution to elicit public acceptance (see Figure 1.3).

Galton's views on eugenics can be summarised in two quotations:

Man is gifted with pity and other kindly feelings; he has also the
power of preventing many kinds of suffering. I conceive it to fall
well within his province to replace Natural Selection by other

processes that are more merciful and not less effective (Galton, 1908).

Eugenics is the scientific study of the biological and social factors which improve or impair the inborn qualities of human beings and of future generations (Galton, 1909).

He was particularly impressed by intellectual eminence, and appreciated excellence in any competitive endeavour. He thought that the leaders of men and the heroes and prophets were set apart from the rest, both psychologically and biologically, and that 'very gifted men are usually of marked individuality, and consequently of a special type', adding:

If a man is gifted with vast intellectual ability, eagerness to work, and power of working, I cannot comprehend how such a man should be suppressed.

On the other hand he found that 'the average man is morally and intellectually an average being', and did not allow that the class of the mediocre might be leavened by men of exceptional potentialities who never showed their capacities; he did not believe in 'mute inglorious Miltons'. The 'lower middle classes' he summed up as mentally and physically litter-scatterers, and once stated that they, and those still lower than them, made up 'the present army of ineffectives which clog progress'.

Along with his low self-esteem, his excessive admiration of leaders of mankind no doubt had its emotional roots in his relationship with his father and with such father-figures as Charles Darwin. But it might additionally be given a rational basis, derived, according to Slater (1960), from his way of looking at normal variation – Galton never made use of the concept of the standard deviation. He thought of people who had been measured in some characteristic as being drawn out in a long line, the man with the lowest measurement to the extreme left and the man with the highest to the extreme right. This distribution, divided conceptually into what we now call percentiles, then led Galton to the misleading idea that there was more variation at the extremes of the distribution than about its middle: the correct view is that variability is a quality of the group as a whole, but that its effects will be more openly manifested at the extremes than near the mode (Slater, 1960).

There is a difficulty with regard to Galton's ideas on the subject of an élite, a subject that determined the way in which he thought a eugenic policy would have to be implemented, as there is no class of human being that can be regarded as constituting an élite. There are individuals who excel, but excel in a limited range of performance, and we have to distinguish between the man and his achievement. Galton, when writing about the senior wrangler at Cambridge whose marks were so much higher than his runner-up, never told us whether he did anything remarkable in his later career. Darwin showed more insight than Galton when he observed, as quoted at the start of this introduction, that the discoverers of undiscovered things were often less clever than those who made no discoveries, and that there are elements of chance and will that intervene between ability and achievement. Small quantitative differences in causes may lead to large, or to qualitative, differences in effects, as may be met with in intelligence testing. High achievement is the reward of the specialist, but it is likely that the specialisation has to be paid for by compensatory deficiencies. As Slater put it, if we are to appraise the value of a mine, we do not take account only of the nuggets of pure gold and forget the ore.

Galton's misconceptions arose because he did all his work without the aid of Mendelism and was working mainly from a social, and little from a biological, point of view. He was right to emphasise the way in which human progress is retarded by sheer stupidity. But the drag on such progress is not the stupidity of the stupid, but the stupidity of the intelligent. For the tragedies of mankind, that army of ineffectives, the intellectual proletariat, are blameless. Through his ignorance of Mendelism Galton was unbiological when he thought that eugenic policies could be achieved by encouraging the fertility of families in which eminent men belonged. The biological worth of an individual depends on his individual make-up, regardless of the constitution of the other members of the family.

In 1878 Charles Darwin wrote:

I have lately been led to reflect a little . . . on the artificial checks [suggested in 'an article by F. Galton, in which he proposes certificates of health and so forth for marriage, and the best should be matched'], but doubt greatly whether such would be advantageous to the world at large at present, however it may be in the distant future (letter to G. A. Gaskell from C. Darwin, dated 15 November, 1878).

Galton's ideas about diplomas of eugenic worth, the encouragement of fertility in 'eminent' families and restrictions on the fertility of the socially incompetent have to be abandoned: in any case they all have the defect of dividing humanity against itself (Slater, 1960). From this it does not necessarily follow, however, that Galton's eugenic ends have also to be dropped, even though the whole complex of ideas comprised under the heading 'eugenics' arouses strong opposition, not least among biologists. In this age of molecular biology the genetic future of mankind still remains to be cared for by inter-disciplinary studies by biosocial scientists, in particular human repro-duction, and development, health and welfare in the widest sense. There must be continued interest in the 'Galtonian' implications of what is *already* being done (for good or ill) in genetic engineering against inherited diseases, and the established practice of prenatal diagnosis and selective abortion.

Galton was the product of his time, holding the views of the affluent part of society to which he belonged. In his zealous advocacy of eugenics he could hardly have anticipated stronger intellectual attack than the mere misrepresentations, and the wrong-headedness of objectors to it, that he recognised in the last chapter of his *Memories* in 1908. He could never have envisaged that his reputation would become so damaged by the misapplication of eugenics, as for instance in the 1930s by the Nazis in Germany when they encouraged 'positive eugenics', with Hitler's cabinet promulgating a eugenic ster-ilisation law, so that the term 'eugenics' took on ugly connotations such as 'genocide' and what is now called 'racism'.

It is unfortunate indeed that the hurtful smear has so continued that in 1992 a leading article in the *British Medical Journal*, consider-ing ethics and the human genome, could contain the following words:

Demand for more conventional health data does not reassure those who are worried about discrimination and the misuse of personal information. Should we add laws against discrimination on the grounds of genetic make-up to those against discrimination on the grounds of race, religion, and gender? The past misappropria-tion of genetics by those with eugenic axes to grind fully justifies these concerns (Carey, 1992).

Here 'eugenics' is being used in a narrow, pejorative sense, embrac-ing the worst sensational distortions of the past and ruling out the present achieved body of acceptable clinical practice.

In his Galton Lecture of 1924 on 'Problems of Race', the anthropologist Grafton Elliot-Smith (1871–1937) dealt with a subject that he thought Galton would have dealt with had he still been alive. Galton's views on race and racial differences reflected the prejudices of his era and his experience, and the only time he wrote on the subject of race was about its improvement. His idea that the human race might be perfected by limiting marriage to the physically and mentally fit was made at a time when the word 'race' did not carry controversial overtones, but signified rather 'mankind'; since then eugenics has too often been purloined away from 'the science of race improvement' towards the connotation of racism. Michael Banton has written in this book a chapter entitled 'Galton's Conception of Race in Historical Perspective' (Chapter 12).

In 1904 Francis Galton founded the Eugenics Record Office, which was housed by University College London. He was its first director, being succeeded by Karl Pearson two years later. The Record Office, in which Galton had endowed a fellowship, in 1907 became the Francis Galton Laboratory for the Study of National Eugenics. When he died in 1911 he left his effects and his estate to the University of London (amounting to £45 000) to found the Galton Professorship of Eugenics, which Pearson then held until 1933 when he retired. The history of the Galton Laboratory is the subject in this book of Chapter 14 by J. S. Jones, its present head.

In his *Essays in Eugenics* (1909), Galton, in summing up the stages in the development of eugenics, described how firstly it 'must be made familiar as an academic question', and secondly that it was 'a practical subject worthy of serious consideration'. Thirdly, he went on, it must be 'introduced into the national conscience, like a new religion'. He then expanded on this theme:

> It has, indeed, strong claims to become an orthodox religious tenet of the future, for Eugenics co-operates with the workings of Nature by securing that humanity shall be represented by the fittest races. What Nature does blindly, slowly, and ruthlessly, man may do providently, quickly, and kindly.

Throughout his career Sir Francis Galton was always a publicist, if not a religious eugenics preacher in his later years, and, looking through his bibliography (which appears later in this book), nowadays he might well be accused of multiple publication. As a member of the board of *Nature* when it was founded in 1869, and as a member of the councils of so many societies, it was easy to get the same article to appear more than once, and of course this was at a time when such practice did not give offence. Part of Galton's need to publish widely must have been his need to educate and convince people of the correctness of what might have appeared, so frequently, to be somewhat cranky views. He was however also interested in teaching, three of Louisa's brothers being headmasters of public schools. He pressed the council of the Royal Geographical Society for their greater involvement in geographical education, so that for sixteen years from 1869 the society conducted examinations and awarded two gold medals yearly for the best performance in physical and in political geography. Besides trying to improve the teaching of geography in schools, he turned his attention to the universities. Galton, as an educationist is the subject in this book of the contribution by W. H. G. Armytage (Chapter 13).

In November 1907 certain members of the Moral Education League met and formed a new society, the Eugenics Education Society, with Sir James Crighton-Browne (1840–1938) as its president. Galton at first promised to help, but then withdrew. Next year, however, the secretary, Montague Crackanthorpe, who was president from 1909 to 1911, persuaded Galton, now aged eighty-six, to give a paper on the origins of eugenics and to become honorary president, a position he retained until his death in 1911, when Leonard Darwin (1850–1943), Charles Darwin's fourth son, became president. The Eugenics Education Society changed its name to the Eugenics Society in 1926; in 1989 its name was changed again, this time to the Galton Institute. A list of the presidents (1908–1992) of all three appears at the end of this book. The *Eugenics Review* was the organ of the two Societies from late 1909 until it closed down in 1984, to be replaced by *Biology and Society*, which is now incorporated into the *Journal of Biosocial Science*.

GALTON'S PERSONALITY

Throughout their married life the Galtons made frequent journeys away from home. Most usually, it seems, the travel was for reasons of both their healths and was abroad, often for many months at a time. But the records are not clear, and it seems impossible to work out to what their ill-health was due. Louisa kept an annual record, with such entries as:

1868. The year began by Frank being too poorly to dine at Mr Crawfurd's. Winter at home, but dined out very little . . . Frank went to his family. Not well in July.

But the entry was sometimes more dramatic, such as in 1874:

On the 14th I broke a blood vessel and was very near dying, but thro' God's mercy, I came back to life and felt so peaceful and happy in my quiet sick room, that it was not a time of misery . . . Frank was ill in December and had Dr A. Clarke. We had a quiet dull Xtmas, no going out and F. had to give up his promised lectures at Newcastle.

Sometimes it seems that Frank suffered nervous symptoms, always attributed to his doing too much.

In 1897, after fourty-four years of married life, and after both of them had been unwell, they went to France, where Louisa, who had become particularly frail, died in Clermont-Ferrand after a bout of diarrhoea (which they all had) and prostrating vomiting. At postmortem examination she was found to have a probable infiltrating cancer of the stomach, or *linitis plastica*. Despite this she is thought to have exaggerated her own ill-health during most of her married life, and had become tiresome with her periodic preparations for death (Forrest, 1974). Invariably she found relief from her complaints when they were abroad and she had her husband's constant companionship. Her bouts of illness and his were often sequential. Her illnesses were an attention-seeking device, and his bodily symptoms appeared whenever he was prevented from working – part of his obsessive personality. Louisa had been a good intellectual companion in their early years, but later was unable to participate in and encourage him in his work. Together they enjoyed literature, other than fiction, but had few other interests in common. He was uninterested in art and actively disliked music. Their political opinions were the same, and Galton never let his religious unbelief obtrude. After Louisa's death Galton returned to Rutland Gate, and his great-niece Eva Biggs came to live with him following a probationary holiday together in Spain and Egypt. In the last ten years of his life he still travelled, but suffered from chronic bronchitis with bronchial asthma, and deafness. As he moved into his eighties he gradually became more feeble, and at the age of eighty-nine, still with a clear mind, he died.

Obsessiveness and depression featured on both sides of Galton's family history, so much so that on the Darwin side Charles Darwin came to see his family as weakened by a poor heredity (Desmond and Moore, 1991). On the Galton side, besides their melancholia the family, including Frank's mother Violetta, were preoccupied with classifying and organising. It was from his Quaker father that Galton derived the puritanical guilt that he implied he lost on reading *Origin of Species*. He twice had a 'nervous breakdown' and 'those who have not suffered from mental breakdown can hardly realise the incapacity it causes' (Galton, 1908). He was liable to nervous symptoms almost throughout his life, and these symptoms, as already mentioned, were usually attributed to doing too much. This is not very convincing as, for instance, the first breakdown occurred when he was reading for mathematical honours in rather a desultory way at a time when he had a very busy social life. Apart from the headaches, inability to concentrate, 'obsessing thoughts' and some degree of depression, there were symptoms of somatic anxiety, such as palpitations and 'giddiness and other maladies prejudicial to mental effort'. But his condition does not seem to have been a simple anxiety state, and the depressive element was not severe in view of its lack of interference with appetite and sleep. The best analysis of Galton's personality and nervous affections is that provided by Slater (1960), although because of the paucity of the available information, he felt the views he arrived at as somewhat inconclusive. Slater (1960) concluded that Galton's trouble was a liability to painful compulsive rumination.

In his *Memories* (1908) Galton does not record anything of his mother's emotional relationship with him, or his feelings for her. The main female influence of his childhood was his much-loved third sister Adèle, who was a semi-invalid. Later he wrote:

My eldest sister was just, my youngest was merciful. When my bread was buttered for me as a child, the former picked out the butter that filled the big holes, the latter did not. Consequently I respected the former, and loved the latter (Galton, 1908).

Apart from Adèle, the dominant influence on his life was his father, at whose death all plans for his career were immediately dropped. He married at the age of thirty-one, and although married for over forty years his wife had no obvious influence in moulding his personality or his views. They remained childless, and Galton, Pearson noted, was alarmed by children and could not find the right words to say to them. It is because of this sort of remark, and perhaps particularly because of his comments about married life, that he has sometimes been thought of as rather a cold person.

Karl Pearson also comments on Galton's incomprehension of the psychology of women, and in his writings and statistics – as several contributors to this book have noted – he often ignored them or disparaged their virtues, as may be shown by quoting from *Inquiries into Human Faculty* (1883):

> I found as a rule that men have more delicate powers of discrimination than women. . . . The tuners of pianofortes are men, and so I understand are the tasters of tea and wine, the sorters of wool. . . . Ladies rarely distinguish the merits of wine at dinner-table, and though custom allows them to preside at the breakfast-table, men think them on the whole to be far from successful makers of tea and coffee.

> The willy-nilly [arbitary] disposition of the female in matters of love is as apparent in the butterfly as in the man.

He noted that women are not only better at carrying heavy loads than man, but are actually not at ease when not carrying them. Any intelligent man, he pointed out, can see this for himself merely by looking at the women passengers on any omnibus. But part of this reaction can be attributed to the age in which he lived and the social class to which he belonged.

Galton's scientific work is characterised by the simplicity and lucidity of his thinking, though one now reads his works despite his style rather than because of it. Often his approach to problems was naïve, without preconceptions, and he attained his ends mainly by ingenuity, logic and abundant patience and capacity for hard work. He attempted to experience the feelings of the insane. 'The method tried was to invest everything I met, whether human, animal or inanimate, with the imaginary attributes of a spy'. The trial was only too successful: by the time he had walked one and a half miles to the cabstand, 'every horse in the stand seemed "watching" him, either with pricked ears, or disguising its espionage' (Galton, 1908). His maxim was whenever you can, count, but there is a strong impression of a compulsive quality, of a need to count for counting's sake. He made a characteristic experiment, or inquiry, into the intensity of boredom in a lecture audience by counting the number of fidgets per man per minute. He found the Royal Geographical Society meetings good hunting ground for fidgets, 'for even there, dull memoirs are occasionally read' (Galton, 1908).

His creativity depended on a lively intelligence, and there were few subjects he tackled where the answer could not be achieved by counting and figuring. Francis Darwin is emphatic that he was never a bore, writing that:

> You might as well call the lightning a bore for explaining that it was going to thunder, or complain of the match for boring the gunpowder as to the proper way of exploding. With Galton's explanations there was a flash of clear words, a delightful smile or gesture' (F. Darwin, 1914).

He was clearly an endearing and lovable man.

I have already mentioned that J. M. Keynes thought that Francis Galton skirted on the right side of eccentricity, but, will now complete the quotation of his remarks of 1937, when he said:

> I wonder how many . . . are well acquainted with his [Galton's] *Art of Travel* – in its peculiar way one of the most entrancing books of adventure ever written. I like to think of him as the white knight, travelling on a donkey through a country infested with lions, wearing socks made that morning out of a flannel strip, with a home-made hairbrush constructed in the light of his own formula, fearful that his donkey's brayings would disclose his progress to the wild beasts, but discovering that if he ties a weight to the donkey's tail all would be well; for with his tail pointing downwards the donkey could not bray. Or, travelling frock-coated through England, with a large railway map torn from Bradshaw in his tail coat pocket to be taken out and pricked with a pin in the station on the map where he found himself, whenever looking out of the window he espied someone exceptionally goodlooking. Let us think on his name with affection and gratitude (M. Keynes, 1984).

References

Bateson, William (1909) *Mendel's Principles of Heredity* (Cambridge: at the University Press).
Carey, N. H. (1992) 'Ethics, money, and the human genome', *British Medical Journal*, vol. 304, pp. 725–6.

Darwin, Charles (1839) *Journal of Researches into the Geology and Natural History of the Various Countries visited by H. M. S. 'Beagle'*, second edn (London: Henry Colburn).

Darwin, Charles (1853) *The Correspondence of Charles Darwin*, vol. 5 (Cambridge University Press, 1989) pp. 149–50.

Darwin, Charles (1859) *On the Origin of Species by Means of Natural Selection or the Preservation of Favoured Races in the Struggle for Life* (London: John Murray).

Darwin, Charles (1871) *The Descent of Man and Selection in Relation to Sex*, 2 vols (London: John Murray).

Darwin, Francis (1914) 'Francis Galton, 1822–1911', *Eugenics Review*, vol. 6, pp. 1–17.

Desmond, A. and J. Moore (1991) *Darwin* (London: Michael Joseph).

Elliot-Smith, G. (1924) 'Problems of Race', reprinted in *Eugenics Review*, vol. 60 (1968) pp. 25–31.

Eysenck, H. J. (1979) *The Structure and Measurement of Intelligence* (New York: Springer-Verlag).

Eysenck, H. J. (1984) 'Intelligence: new wine in old bottles', in C. J. Turner and H. B. Miles (eds), *The Biology of Human Intelligence* (London: The Eugenics Society).

Faulds, Henry (1880) 'On the skin furrows of the hand', *Nature*, vol. 22, p. 605.

Fisher, R. A. (1936) 'Has Mendel's work been rediscovered?' *Ann. Sci.*, vol. 1, pp. 115–37.

Forrest, D. W. (1974) *Francis Galton – The Life and Work of a Victorian Genius* (London: Paul Elek).

Galton, Francis (1852) 'Recent expedition into the interior of South-Western Africa', *Journal of the Royal Geographical Society*, vol. 22, pp. 140–63.

Galton, Francis (1855) *The Art of Travel; or, Shifts and Contrivances Available in Wild Countries* (London: John Murray).

Galton, Francis (1863) 'A development of the theory of cyclones (anti-cyclones)', *Proceedings of the Royal Society*, vol. 12, pp. 385–6.

Galton, Francis (1863) *Meteorographica, or Methods of Mapping the Weather* (London and Cambridge: Macmillan).

Galton, Francis (1865) 'Hereditary talent and character', *Macmillan's Magazine*, vol. 12, pp. 157–66, 318–27.

Galton, Francis (1869) *Hereditary Genius* (London: Macmillan).

Galton, Francis (1872) 'Statistical inquiries into the efficacy of prayer', *Fortnightly Review*, vol. 12, pp. 125–35.

Galton, Francis (1874) *English Men of Science: their Nature and Nurture* (London: Macmillan).

Galton, Francis (1874) 'Notes on the Marlborough School statistics', *Journal of the Anthropological Institute*, vol. 4, pp. 130–5.

Galton, Francis (1875) 'The history of twins as a criterion of the relative powers of nature and nurture', *Fraser's Magazine*, vol. 12, pp. 566–76.

Galton, Francis (1877) 'Typical laws of heredity', *Nature*, vol. 15, pp. 492–5, 512–4, 532–3.

Galton, Francis (1883) *Inquiries into Human Faculty and its Development*, (London: Macmillan).

Galton, Francis (1885) 'Regression towards mediocrity in hereditary stature', *Journal of the Anthropological Institute*, vol. 15, pp. 246–63.

Galton, Francis (1888a) 'Co-relations and their measurement, chiefly from anthropometric data', *Proceedings of the Royal Society*, vol. 45, pp. 135–45.

Galton, Francis (1888b) 'Personal identification and description', *Proceedings of the Royal Institute*, vol. 12, pp. 346–60.

Galton, Francis (1889) *Natural Inheritance* (London: Macmillan).

Galton, Francis (1892) *Finger Prints* (London: Macmillan).

Galton, Francis (1908) *Memories of my Life* (London: Methuen).

Galton, Francis (1909) *Essays in Eugenics* (London: Eugenics Education Society).

Herschel, Sir William (1880) 'Skin furrows of the hand', *Nature*, vol. 23, p. 76.

Keith, Arthur (1920) 'Galton's Place among Anthropologists', reprinted in *Eugenics Review*, vol. 60 (1968) pp. 12–24.

Keynes, J. M. (1937) 'Some Economic Consequencies of a Declining Population', *Eugenics Review*, vol. 29, pp. 13–17.

Keynes, Milo (1984) 'Centenary of the birth of John Maynard Keynes', *Biology and Society*, vol. 1, pp. 10–13.

Mendel, Gregor (1865) 'Versuche über Pflanzen-Hybriden', *Verhandlungen des naturforschenden Vereines in Brünn*, vol. 4, pp. 3–47; English translation in *Journal of the Royal Horticultural Society*, vol. 26 (1901).

Middleton, Dorothy (1982) *Sir Francis Galton 1822–1911 – Jubilee Memoir of the Eugenics Society* (privately printed by the Eugenics Society), pp. 38.

Pearson, Karl (1914, 1924 and 1930) *The Life, Letters and Labours of Francis Galton*, 3 vols (Cambridge at the University Press).

Royal Geographical Society (1854) 'Hints to travellers', *Journal of the Royal Geographical Society*, vol. 24, pp. 1–13.

Shaw, G. B. (1905) [Comments on] F. Galton. 'Eugenics. Its definition, scope and aims', *Sociological Papers*, vol. 1, pp. 45–50, 78–9.

Skanes, G. R. (1984) 'Eighteen eighty-four', *Canadian Psychology*, vol. 25, pp. 258–68.

Slater, Eliot (1960) 'Galton's Heritage', *Eugenics Review*, vol. 52, pp. 91–103.

2 The Historical Galton

W. F. Bynum

Francis Galton shared much with those fellow Victorians who created of our modern world view. Like his cousin Charles Darwin, Alfred Russel Wallace, Joseph Dalton Hooker and Thomas Henry Huxley, he spent some of his formative years in foreign travel to exotic parts. Like Darwin, Wallace and Herbert Spencer, he never held an academic post. Like Darwin, Huxley and Hooker, he received the rudiments of medical training. Like William Crookes, Wallace and Huxley, he attended seances and flirted with spiritualism; like John Tyndall, Darwin and Huxley, he rejected the realities of the unseen world in favour of the here and now, and the future of man on earth. Like Wallace, Hooker and Spencer, his active, intellectual life spanned the whole of Victoria's reign and beyond. Like the good Victorians they were, all these men suffered from all manner of niggling complaints, had regular breakdowns in health and wheezed and cosseted themselves to an average age of eighty-two and a half.

Like Tyndall and Huxley, and, in more subtle ways, Wallace and Spencer, Galton's life was irrevocably changed in 1859 with the publication of Darwin's *Origin of Species*. Like all of these scientists, he published his most important monographs with mainline publishers in reasonable print-runs, and had them seriously reviewed in major general periodicals with large circulations among the Victorian reading public (Young, 1985; Ellegård, 1958). Like his other long-lived compatriots, he saw, but did not benefit from, the solid beginnings of professional science in Britain; Galton in fact was one of the benefactors of professional science in this country (Alter, 1987). His own career and contributions lay completely outside any notion of the professional, within an amateur tradition that did not recognise disciplinary boundaries and, in its fullest realisations, including Galton's, was all the richer for it (Morrell and Thackray, 1981). The breadths of his interests and intellectual curiosity are only partially covered by this volume.

Nevertheless Galton remains something of an historical enigma, especially in relation to the most obvious point of comparison, Charles Darwin. Historians continue to lavish minute attention on Darwin, his intellectual development, his ideas and influences, his health and personality. The magisterial edition of his collected *Correspondence* is well in train and bids fair to keep a team of scholars in work until at least the end of the century. A new biography of Darwin is published about every year; indeed there have been three within the past year, each substantial and offering a different interpretation of the man whom the latest biographers, Adrian Desmond and James Moore, have dubbed the Squire of Downe (Bowlby, 1990; Bowler, 1990; Desmond and Moore, 1991).

In contrast Galton has been at best compartmentalised, at worst virtually overlooked. Thus the entry on him in the supposedly authoritative *Dictionary of Scientific Biography* (1972) is almost perfunctory, a good deal shorter than that on Spencer, less than a quarter the length of Wallace's, and a tenth that of his cousin. He rates no mention at all in Walter Houghton's classic *Victorian Frame of Mind* (1957), but this might be partially explained by Houghton's cut-off date of 1870, too early for much of Galton's most enduring work. The massive survey of *Edwardian England* does little better, with Galton trailing Darwin by two citations to one (Nowell-Smith, 1964). Frank Turner's study of late Victorian scientific naturalism, appropriately entitled *Between Science and Religion* (1974), could have made much of Galton, who after all scandalised Victorian sensibilities by his attempts to show statistically that prayer is objectively inefficacious (Galton, 1872). Instead Galton played only a walk-on part in Turner's drama.

Historians of science, or scientists writing about the history of their disciplines, have done somewhat better, even if they considered only discrete aspects of Galton's array of interests (Cowan, 1972a, 1972b, 1977). Historians of the eugenics movement, such as Geoffrey Searle (1976), Donald Mackenzie (1976) and Daniel Kevles (1986), have of necessity made Galton a prominent figure in their monographs. Kevles in particular offered an astute portrait of Galton's personality before analysing the influence of his ideas in Britain and the United States. Mackenzie was rather more interested in viewing Galton as a spokesman for a set of middle-class anxieties and aspirations.

The hereditarian dimensions of Galton's work have also been placed within the context of the history of genetics. Thus Robert Olby (1985) and Peter Bowler (1989) have offered knowledgeable accounts of Galton's important role in the development of what is sometimes called 'hard heredity', in contrast to the softer, environmentally sensitive theories which were more dominant in Galton's time. No biographer of Darwin fails to mention the experimental critique which Galton offered of Darwin's theory of pangenesis, though some of them fail to notice that one of Galton's earliest writings on inheritance flirted with a theory very close to that of pangenesis.

Nor has Galton been omitted from accounts of the post-*Origin* history of evolutionary theory. Scholars like Bowler (1983) have noted Galton's preoccupation with Darwinianism during the period of its 'eclipse' in the late nineteenth century, and the 'biometric-Mendelian' debates of the early twentieth century have been examined not simply for the social values of the antagonists, but also as providing the conceptual background for the 'neo-Darwinian' evolutionary synthesis effected by Ronald Fisher, J. B. S. Haldane, Sewall Wright and others from the 1920s (Norton, 1973; Provine, 1971, 1986).

The statisticians have not forgotten Galton (Mackenzie, 1982). Stephen Stigler (1986) describes Galton as 'a romantic figure in the history of statistics', and Theodore Porter (1986) assigns him a leading role in the 'rise of statistical thinking'. Neither Karl Pearson, Galton's most devoted disciple, nor Pearson's son Egon, a distinguished statistician in his own right, were backward in singing Galton's praises in their own historical writings (Pearson and Kendall, 1970).

Others have recognised Galton's innovations in a variety of additional fields of enquiry. Hearnshaw (1964) and Fancher (1990) have placed him in the development of psychology, especially in Britain; Tanner (1981) has explicated his importance for the measurement of human growth and form; students of the history of anthropology, criminology, the use of photography for scientific purposes, travel and geographical exploration, cartography, fingerprinting, meteorology and inventions have also paid him his due. He has been dubbed a 'genius' by one biographer (Forrest, 1974) and a 'dilettante and maverick' by an historian of evolutionary biology (Mayr, 1982).

If there have been many who would analyse parts of the man and his work, there has been a surprising reluctance to synthesise the man as a whole. In fact there are only four substantial instances, including the one which Galton himself left us in the form of his autobiography, published in 1908 when he was eighty-six (Galton, 1908). Galton's assessment of himself was followed by Pearson's monumental *Life, Letters and Labours of Francis Galton* (1914–1930), the first volume of which appeared in 1914, when Pearson was still at the height of his powers, with a further volume in 1924 and the final two in 1930, when Pearson was nearing the end of his own life. C. P. Blacker, long-time general secretary of the Eugenics Society, devoted about half of his monograph on *Eugenics: Galton and After* (1952) to Galton himself, and though he made the man more accessible, he rarely went beyond Galton's *Memories* or Pearson's *Life* for his information. In 1974 D. W. Forrest published a sympathetic biography which made reasonable use of the extensive collection of Galton's manuscripts in the library of University College London; of material held by what was then called the Eugenics Society, which is now housed in the Library and the Contemporary Medical Archives Centre of the Wellcome Institute; and of the archives of the Royal Geographical Society. However Forrest apparently made little attempt to assess the archives of other societies with which Galton was connected, nor the private papers of some of his many correspondents. Perhaps he took Pearson at his word, when Pearson complained that his own searches had not yielded much beyond the admittedly large cache of material which had come to the Galton Laboratories at University College at the time of Galton's death, and the additional body of family papers then still in the hands of members of the Galton, Barber and Darwin families (Merrington and Golden, 1976).

A new biography of Galton is a desideratum, and any future biographer will have more reason to praise than to rail at Pearson, although Pearson himself admitted that in the later volumes, as his eyesight and probably his energy began to fail, he had trouble reading some of the handwriting. And even Galton might have blushed at the saintly portrait which Pearson painted. Pearson's *Life of Galton* was clearly a labour of love, although sometimes as revealing of its author as its subject. It was also, even as it appeared, something of a fossil. Two or three volumes were standard fare for a Victorian *Life and Letters*, usually edited by a devoted wife or son soon after the distinguished subject's death. Darwin and Huxley each got three; Charles Lyell, Hooker and many other scientific luminaries two; Wallace and Spencer only one, although these had been preceded by late-life double-decker autobiographies.

By the early twentieth century, when Pearson was writing, tastes were beginning to change and the monumental format, massive detail, not simply about Galton but about dozens of his ancestors, lavish production standards, and initial publication on the eve of the First World War make it understandable that, in Pearson's words, the first volume 'met with few readers, and failed to repay the cost of production'. Later volumes were possible only because subsidies were found from private benefactors. Nevertheless the successive volumes of the biography attracted respectful, even enthusiastic, reviews. Raymond Pearl (1925, 1931) sang its praises in *Science*; one reviewer compared Pearson to Boswell; and the anonymous reviewers for *Nature* were equally fulsome in their admiration for Pearson's achievement and claimed for Galton the completion of Darwin's anthropological agenda: Darwin, the *Nature* reviewer concluded, had uncovered mankind evolved; Galton had discovered mankind evolving (Anon, 1925; Anon, 1931).

For all the detail, and the hundreds of Galton's letters which Pearson included, it is hard to discover the man in the *Life*. Galton, it has to be said, was not a brilliant letter writer. Some of the early letters are full of interest, for example those describing his medical training at the Birmingham General Hospital and at King's College, London. In these Galton not only described some of the emotional experiences of a young man – a teenager in fact – confronting the realities of disease, death and decay, he also suggested that, had he chosen to follow a medical career, he might have pioneered a mathematically more rigorous approach to therapeutics. Even as a relatively untutored medical student he was aware of the difficulties in assessing the efficacy of therapeutic interventions in the clinical setting. Some of Galton's later correspondence contains valuable scientific insights and important clues to his working methods. In between, his travel letters – the early trip to the Continent, when he took himself off, aged eighteen, to Constantinople and the Middle East, the later trips to Egypt and Syria, and finally to South-West Africa – reveal but little of his inner thoughts or intellectual development. Or if he did reveal himself, especially his growing intolerance of all religious beliefs combined with the attrition of his own never very intense religious sentiments, these letters have not survived.

Nor was his chief biographer, Karl Pearson, the most subtle of psychologists. His extensive physical and psychological analyses of Galton were rather clumsily done: a crude listing of traits with long discussions of which ancestor or ancestors had passed them on to him. When Galton on occasion seemed to transcend his ancestry, or to change his own habits by what in common language would be called an act of will, Pearson merely retreated to the proposition that this capacity, too, is inherited. Forrest remarked that in 2000 pages of quarto text Pearson had buried the man beneath the monument. But it was not all Pearson's fault, for Galton displayed little of his inner self even in his own autobiography. *Memories of my Life* is a remarkable volume, with a freshness and buoyancy unusual among octogenarians. Darwin, Huxley and Spencer grew old. Not so Wallace and Galton, who remained evergreen. Pearson characterised Galton as never growing old, only wise. Perhaps it was that wisdom, combined with a genuine dislike of controversy, that encouraged Galton to place a mask between himself and his reader. His was no autobiography in the tradition of St. Augustine, Rousseau or even Darwin. It contained little self-analysis, little personal or domestic detail, little of the inner man, even by way of an *Apologia*. What the volume does reveal is Galton's sociability and capacity for making a variety of friends and acquaintances among the Victorian middle classes, and it invites some future biographer to place him in a broader social context, to pay more attention to Galton as others saw him. As he implied in the preface, most of the friends that he had outlived were to be found in the *Dictionary of National Biography*. He numbered among his friends and acquaintances judges and politicians, explorers and doctors, scientists and social reformers, dons and clergymen. By both birth and inclination, Galton was part of that remarkable group which Nöel Annan (1955) has dubbed the intellectual aristocracy, and which has played so important a part in the cultural, academic and scientific life of Britain during the past two centuries. Galton's own extended family included not just the Galtons and the Darwins, but the Savilles, Colliers and Barclays. He married into the Butlers, which family produced bishops, politicians, colonial administrators and two masters of Trinity College, Cambridge, where Galton himself had had a busy but only modestly successful undergraduate career. It was probably the ordinariness of his academic achievements at Trinity which made him doubly pleased to receive, late in life, an honorary fellowship from his old college, at a time when his brother-in-law, Henry Montagu Butler, was its master.

Marrying into the Butler family was obviously a source of pride and satisfaction to Galton, not least for the fact that, like his own blood relations, the Butlers seemed living proof that talent runs in families. Nevertheless it is not simple prurience that makes one wish to know more about Galton's relationship with his wife, Louisa Butler. He spoke of her fondly enough in his *Memories*, and the letter he wrote to his favourite sister, Emma (whom he dubbed 'my loving and beloved correspondent', on Louisa's sudden, though not completely unexpected, death in 1897, shows a certain depth of feeling; he had, after all, been married to her for forty-four years. For all that she remains a shadowy figure, and the reader of Galton's *Memories* would not even know her Christian name. She was introduced in the autobiography when, without reference to any courtship, Galton

> became engaged to my future wife, the daughter of the Very Rev. George Butler, Dean of Peterborough, who had been Head-master of Harrow during many years. My wife had three sisters and four brothers, the latter all highly distinguished for scholastic and administrative ability (p. 154).

Subsequently Louisa appears only as 'my wife', and in the index as 'Mrs Francis Galton (my wife)'.

'He has no children', says the grieving MacDuff of Macbeth; one of the many puzzles of Shakespeare's great play of fruitfulness and barrenness since Lady Macbeth had already professed to giving suck. Like Macbeth, the father of eugenics had no children, though there is no evidence of why this was so, or of how it affected Galton in his preoccupation with family and fortune, talent and success, or his concern that individuals just like himself should as a social duty have large numbers of offspring. How deeply his childlessness affected him we shall never know, for he rarely wore his heart upon his sleeve.

Nor do his *Memories* reveal much about his religious sentiments. We know that Darwin's candid comments in his own autobiography about his religious beliefs pained some members of his family and were edited out of the version published by Darwin's son Francis while Darwin's wife, Emma, was still alive (Darwin, 1958). Clergy-men regularly appear as Galton's friends, acquaintances and travel-ling companions in his own autobiography, even after 1872 when he published his infamous 'Statistical inquiries into the efficacy of prayer' in the *Fortnightly Review* (Galton, 1872). Galton never men-tioned that paper in his *Memories*, nor, interestingly, did Louisa in her private record of their domestic life of 1872. Galton wrote of the church service and 'appropriate hymns' following Louisa's death, and there is the possibility that, like Emma Darwin, Louisa Galton found her husband's agnosticism difficult to comprehend.

Despite the blandness of the *Memories*, Galton had never shied from public statements on the historical origins and psychological functions of religion. All of his biographers agree that for the last fifty years of his life Galton was an agnostic, albeit probably what his niece described as a 'religious agnostic'. By that was meant that he took religion seriously and recognised its social and psychological value, even while denying the transcendent origin or unique claims to truth of any religion. What took the place for him of any formal religious creed was, quite simply, eugenics. Having lost his sense of the transcendent, Galton placed his faith in the planned, rational and scientific breeding of mankind. He was, according to Pearson, a 'Freethinking pantheist, who desired to reach a socialistic state by breeding supermen for intellect' (vol. II, p. 119).

Pearson may have been describing himself more accurately than his mentor, for in truth Galton was never caught up in the socialist fervour which so dominated Pearson in the 1870s and 80s (Norton, 1978). There was a generation gap between Galton and Pearson, and for all Galton's intellectual flexibility he remained from middle-age a socially conventional Victorian gentleman, frequenting the Athenaeum, active in the British Association for the Advancement of Science, eschewing controversy, wintering in the South of France. Like Richard Burton and Henry Stanley, Galton had felt the lure of Africa, had experienced the reality of the exotic, had, so he tells us, confronted the hospitalities of foreign parts, where wives were freely offered as tokens of peace and friendship.

But if Galton had gone to the Middle East or Africa, like more than a few of his contemporaries, to experience its exotic and carnal delights, he kept his dalliances well hidden. Forrest has uncovered one letter (to, not from, Galton) which hints that Galton may have spent the traditional night with Venus, with months of Mercury to follow, but that single letter is vague and hardly conclusive, even reading it with all the awareness of Victorian euphemism (Forrest, 1974, p. 33). In the absence of better evidence we are forced to conclude that Galton's youthful *wanderlust* was merely the expression of a trait that never left him: an incapacity to enjoy the tame pursuits of English country life for more than short periods of time, combined with a desire to vasciilate between the rational pursuits of the metropolis and an itch for foreign travel. After his final exotic excursion to South-West Africa he became uncertain of his health, and, especially after his marriage, seems to have been content with the foreignness of Continental Spas, Switzerland and the Alps, or the South of France. He returned to Egypt once after his wife's death. This was not a trajectory followed by many of his fellow Victorian explorers, and it was surprisingly tame, perhaps, in one like Galton who made his scientific reputation as a geographer of the unknown and continued to be active in the affairs of the Royal Geographical Society. But, then, Galton never was quite predictable, and was perhaps actually unconventional in the way he adjusted to London life after his years in Darkest Africa.

Galton remains elusive in yet another way, for, despite the fact that he had a wide circle of friends and was a good scientific committee man, he remained in his creative life a curiously solitary figure; more solitary, in a way, than Darwin, cosseted away in darkest Downe yet vitally dependent on Sir Rowland Hill's penny post. Of course Galton enlisted much help in his various projects: Fellows of the Royal Society were persuaded to fill out questionnaires about themselves and their ancestors for him, and friends from all round the country were engaged to raise sweet peas and measure their size on his behalf. For all that, it is sometimes difficult to place him in an ordinary scientific tradition. This is partly because of the sheer breadth of his interests and partly because he was never much of a reader. He seems to have published only two book reviews in his entire life. His own library, meagre and unimportant to him, consisted mostly of books and articles which people happened to have sent to him. He apparently never seriously read Laplace or Poisson, Gauss or Quetelet, despite his own place within the statistical tradition they had created. There simply could not be a Galton industry of quite the same character as the Darwin industry, where marginalia and underlinings of Darwin's books are eagerly scrutinised. Darwin peppered his writings with discussions of, and footnotes to, other people's work; not so Galton. This does not mean that Galton was less generous than Darwin, merely that Galton's major ideas came so much from within himself. Occasionally, as in the case of his volume on fingerprints, he became involved in priority disputes, an episode that pained him, as yet another instance of the controversy he never sought and rarely engaged in (Galton, 1965). 'Do not resent criticism and never answer it', was the philosophy he adopted (Pearson, 1914–30, vol. IIIA, p. 399). Pearson was of the opinion that Galton could have had more impact had he responded more systematically to his critics, particularly those who were not convinced by the hereditarian cast of so much of his later research and writing. If so, perhaps Galton himself was responsible for his relative historical anonymity.

Galton, then, was not much of a reader. Nevertheless one book left its indelible mark upon him, as evidenced in his *Memories*, forty-nine years after the event:

The publication in 1859 of the *Origin of Species* by Charles Darwin made a marked epoch in my own mental development, as it did in that of human thought generally. Its effect was to demolish a multitude of dogmatic barriers by a single stroke, and to arouse a spirit of rebellion against all ancient authorities whose positive and unauthenticated statements were contradicted by modern science (Galton, 1908, p. 287).

On 9 December 1859 he had written to Darwin that

I have laid it [*Origin of Species*] down in the full enjoyment of a feeling that one rarely experiences after boyish days, of having been initiated into an entirely new province of knowledge which, nevertheless, connects itself with other things in a thousand ways (Pearson, 1914–30).

Neither the initial reaction nor the recollection overstated the case. Of all Victorians there is no better example than Galton as one whose intellectual life utterly hinged on Darwin's work. Spencer and Wallace were evolutionists before *Origin of Species*; Lyell could never stomach the full ramifications of the book's thesis; Hooker fell into line without seriously altering the course of his career; Huxley had his career transformed but never quite assimilated the nuances of Darwin's proposed mechanism for speciation; but *Origin of Species* turned Francis Galton from a Jack of several trades to a master of a great many more; from something of a clever dilettante to a major man of science; from a student of the earth to an anthropologist in the most comprehensive sense of the term. Unlike Huxley, who became a noisy evolutionist, Galton became a quiet one. But without the sense of process which Darwin's work provided the whole thrust of his research after the mid-1860s is unimaginable. The legacy of that research is the subject of this volume.

Acknowledgements

Research expenses have been generously met by the Wellcome Trust. The staff of the Library of the Wellcome Institute have, as always, been efficient and helpful. Caroline Prentice provided bibliographical assistance at short notice, and Sally Bragg kept track of everything.

References

Alter, P. (1987) *The Reluctant Patron: Science and the State in Britain, 1950–1920*, translated by A. Davis (Oxford: Berg).

Annan, N. (1955) 'The Intellectual Aristocracy', in *Studies in Social History: a tribute to G. M. Trevelyan*, edited by J. H. Plumb (London, New York, Toronto: Longmans, Green & Co.)

Anon. (1925) review of Vols I and II of Pearson's *Life of Galton*, *Nature*, vol. 115, pp. 631–4.

Anon. (1931) review of Vols III and IIIA, Pearson's *Life of Galton*, *Nature*, vol. 127, pp. 432–4.

Blacker, C. (1952) *Eugenics: Galton and After* (London: Duckworth).

Bowlby, J. (1990) *Charles Darwin. A Biography* (London: Hutchinson).

Bowler, P. (1983) *The Eclipse of Darwinism: Anti-Darwinian Evolution Theories in the Decades around 1900* (Baltimore: Johns Hopkins University Press).

Bowler, P. (1989) *The Mendelian Revolution* (London: The Athlone Press).

Bowler, P. (1990) *Charles Darwin. The Man and His Influence* (Oxford: Basic Blackwell).

Cowan, R. (1972a) 'Francis Galton's Contributions to Genetics', *Journal of the History of Biology*, vol. 5, pp. 389–412.

Cowan, R. (1972b) 'Francis Galton's Statistical Ideas: The Influence of Eugenics', *Isis*, vol. 63, pp. 509–28.

Cowan, R. (1977) 'Nature and Nurture: The Interplay of Biology and Politics in the Work of Francis Galton', *Studies in the History of Biology*, vol. 1, pp. 133–208.

Darwin, C. (1958) *The Autobiography of Charles Darwin*, edited by Nora Barlow (London: Collins).

Desmond, A. and Moore, J. (1991) *Darwin* (London: Michael Joseph).

Dictionary of Scientific Biography (1972) *Sir Francis Galton*, vol. 5, pp. 265–7 (New York: Charles Scribners Sons).

Ellegård, A. (1958) *Darwin and the General Reader: The Reception of Darwin's Theory of Evolution in the British Periodical Press* (Göteburg: Acta Universitatis Gothenburgensis).

Fancher, R. (1990) *Pioneers of Psychology*, 2nd edn (New York: W.W. Norton).

Forrest, D. (1974) *Francis Galton: The Life and Work of a Victorian Genius* (London: Paul Elek).

Galton, F. (1872) 'Statistical Inquiries into the Efficacy of Prayer', *Fortnightly Review*, vol. 12, pp. 125–35.

Galton, F. (1908) *Memories of My Life* (London: Methuen).

Galton, F. (1965) *Finger Prints*; Preface by H. Cummins (New York: Da Capo Press).

Hearnshaw, L. (1964) *A Short History of British Psychology* (London: Methuen).

Houghton, W. (1957) *The Victorian Frame of Mind, 1830–1870* (New Haven: Yale University Press).

Kevles, D. (1986) *In the Name of Eugenics: Genetics and the Uses of Human Heredity* (Harmondsworth: Penguin).

Mackenzie, D. (1976) 'Eugenics in Britain', *Social Studies in Science*, vol. 6, pp. 499–532.

Mackenzie, D. (1982) *Statistics in Britain, 1865–1930: The Social Construction of Scientific Knowledge* (Edinburgh University Press).

Mayr, E. (1982) *The Growth of Biological Thought* (Cambridge, Mass and London: Harvard University Press).

Merrington, M. and J. Golden (1976) *A List of the Papers and Correspondence of Sir Francis Galton (1832–1911) held in the Manuscripts Room, the Library, University College London* (London: Galton Laboratory).

Morrell, J. and A. Thackray (1981) *Gentlemen of Science: Early Years of the British Association for the Advancement of Science* (Oxford: Clarendon Press).

Norton, B. (1973) 'The Biometric Defense of Darwinism', *Journal of the History of Biology*, vol. 6, pp. 283–316.

Norton, B. (1978) 'Karl Pearson and Statistics: The Social Origins of Scientific Innovation', *Social Studies of Science*, vol. 8, pp. 3–34.

Nowell-Smith, S. (ed.) (1964) *Edwardian England, 1901–1914* (London: Oxford University Press).

Olby, R. (1985) *The Origins of Mendelism*, 2nd edn (University of Chicago Press).

Pearl, R. (1925) review of Vols I and II of Pearson's *Life of Galton*, *Science*, vol. 61, pp. 209–12.

Pearl, R. (1931) review of Vols IIIA and IIIB of Pearson's *Life of Galton*, *Science*, vol. 73, pp. 238–240.

Pearson, E. and M. Kendall (eds) (1970) *Studies in the History of Statistics and Probability* (London: Charles Griffin).

Pearson, K. (1914–30) *The Life, Letters and Labours of Francis Galton*, 3 vols in 4 parts (Cambridge University Press).

Porter, T. (1986) *The Rise of Statistical Thinking* (Princeton University Press).

Provine, W. (1971) *The Origins of Theoretical Population Genetics* (University of Chicago Press).

Provine, W. (1986) *Sewell Wright and Evolutionary Biology* (Chicago and London: University of Chicago Press).

Searle, G. (1976) *Eugenics and Politics in Britain: 1900–1914* (Leiden: Noordhoff).

Stigler, S. (1986) *The History of Statistics: The Measurement of Uncertainty before 1900* (Cambridge, Mass and London: Harvard University Press).

Tanner, J. (1981) *A History of the Study of Human Growth* (Cambridge University Press).

Turner, F. (1974) *Between Science and Religion: The Reaction to Scientific Naturalism in Late Victorian England* (New Haven and London: Yale University Press).

Young, R. (1985) *Darwin's Metaphor: Nature's Place in Victorian Culture* (Cambridge University Press).

3 Francis Galton: Travel and Geography

Dorothy Middleton

Not long before he died Galton wrote to a nephew: 'My version of the saying "Whom the Gods love, die young", is that they still *feel* young however old in years they may be. I think, from that point of view, that the Gods must have some *tendresse* to me, because although I am $87\frac{1}{2}$ years old, I hardly feel myself to be even matured'. It was this perpetual spring of youth, stimulating and sustaining an insatiable curiosity in the world about him, that made Galton such a splendid traveller and his experiences so entertaining to read about. It does not matter that his African safari of 1850–2 is not in itself of great importance in the history of African exploration; what matters is Galton's description of it all.

Namibia is in fact an arid land, presenting no exciting challenges, no mysterious river sources, no hidden mountains, and Galton himself neither ranks, nor would have wished to rank, with the heroes of his day – with Livingstone and Burton, Baker and Speke. He had no wish to rival either his predecessors or his contemporaries, competitive to a man, willing indeed to die in pursuit of the goal, to be *first* – Mungo Park drowned in the Niger, Gordon Laing was murdered at the gates of Timbuctu. Galton's was a happier ambition: 'If you have health, a great craving for adventure, at least a moderate income and set your heart on a definite object, then travel by all means' was his advice (Galton, 1872, p. 1). And: 'Interest yourself chiefly in the progress of your journey and do not look forward to its end with eagerness. It is better to think of a return to civil life not as an end to hardship and a haven from ill, but as a close to an adventurous and pleasant life' (Galton, 1872, p. 4).

Galton's first foray overseas was in 1840 when, after two gruelling years as a medical student, 'A passion for travel took me as if I had been a migratory bird' (Galton, 1908, p. 48). His father had arranged a course of lectures on chemistry for him in Germany, but that was too much like work and Frank played truant, sailing down the Danube, across the Black Sea to Troy and Smyrna and the romantic lands of antiquity which our ancestors called Asia Minor. Everything was grist to the mill of his teeming brain – it took nearly 100 horses to haul barges through the Iron Gates, he noted; a shoal of watersnakes, 'A sight upon which a horrible nightmare might have been founded', stirred his imagination (Galton, 1908, p. 51). He could even find some amusement in the tedium of quarantine, imposed on travellers coming from Turkey. Young men were allowed to cut short the time by stripping naked and swimming across a stretch of water to where old-clothes merchants waited with second-hand garments in exchange for those which had been discarded. 'The clothes were thin, and the trousers were made of a sort of calico and deficient in the fashionable cut of my old ones, but I did not care' (Galton, 1908, p. 55).

Five years later, coming into money on his father's death, he abandoned his medical career and was off again. This time he went to Egypt and some way up the Nile, interesting himself in the habits of camels and making friends with the Abyssinian traveller Mansfield Parkyns, who was living in Khartoum. It was another five years before the 'migratory bird' took flight again and this time it was to be a genuine African exploration.

Galton was ready for his adventure at a time of great acceleration in the progress of geographical discovery. It was a movement sponsored and encouraged by the Royal Geographical Society, founded in 1830 and, at the time of Galton's election in 1849, concentrating on Africa. The RGS was probably pleased with its new Fellow, who was healthy, clever, rich and independent (he was paying his own way). And it would have done him no harm that he belonged to the social class from which, at that time, the learned societies of London mostly recruited their governing bodies. They were socially, in fact, rather a closed shop, to be penetrated only by exceptional characters from a different world who had distinguished themselves in the field – Livingstone, for instance, and later Joseph Thomson.

Galton was soon discussing his plans with the geographical establishment. He had at first thought of following in the footsteps of Livingstone across the Kalahari to Lake Ngami, but on arrival at the Cape in the summer of 1850 he was to find the way to the north barred by conflict between land-hungry Boers and the local Bantu. So, accompanied by Charles James Andersson, an Anglo–Swedish naturalist engaged as an assistant for the trip, he sailed up the south-west coast of Africa to Walvis Bay with a view to travelling eastwards to Ngami. In fact Ngami was abandoned and instead they carried out a partial exploration of what we now know as Namibia, but which Galton called 'Damaraland', the territory of the Damaras (or Herreros) north of the Swakop river (Figure 3.1). To the south lay the lands of the Namaqua Hottentots whose chief sport and occupation was raiding their northern neighbours and stealing their cattle. Between, along the river, German missionaries were at work and in constant danger from both sides. Since Galton and Andersson meant to base themselves with the friendly Germans, no advance was possible unless their rear could be secured; Galton determined to take a firm line and one of his best stories is of how he tackled Jonker Afrikaner, the powerful Hottentot chief who was responsible for most of the trouble. Dressing himself in full English hunting rig, he mounted his ox, Ceylon, and galloped into the chief's kraal demanding that he keep the peace on pain of punitive action by the government at the Cape. Galton could hardly have called upon so distant an authority, but either his masterful manner or his extraordinary appearance did the trick and things stayed quiet while he was in the country. 'Oxen are not bad leapers' he reflected, as Ceylon cleared the ditch protecting the hostile camp, 'If you give them time' (Galton, 1889, pp. 69–70).

They made their way north in the traditional South African style by ox-wagon, accompanied by a drove of cattle to serve partly as what Galton called an 'itinerant larder', partly to reinforce the wagon teams. Mounted on Ceylon, Galton rode alongside the herd, a hundred strong, amusing himself by observing their movements and habits. At night he camped where he could watch the herd 'in a group round the fire, chewing the cud with their large eyes glaring in the light, apparently thinking' (Galton, 1889, p. 67). They carried few supplies, depending on their 'itinerant larder' and what they could shoot for the pot, and on what water they might find by the way. They travelled through Damaraland into the country of the Ovampo people and nearly as far as the Cunene river which marks the border of Angola.

Galton was chiefly interested in mapping the terrain, Andersson in natural history. In Galton's opinion, survey was the most important duty of an explorer. 'A traveller who sets about mapping his route should always bear in mind the object for which he maps' was his advice, 'The one to afford a quick guide to future travellers, the other to give a good general idea of the country' (U. C. L. Library: Galton Paper no. 97). He was, at that time perhaps, hardly an anthropologist as we understand the term, but he was interested in the local people and describes them accurately enough. He took particular note of the cattle-owning Damaras who 'have a great respect, almost reverence

48

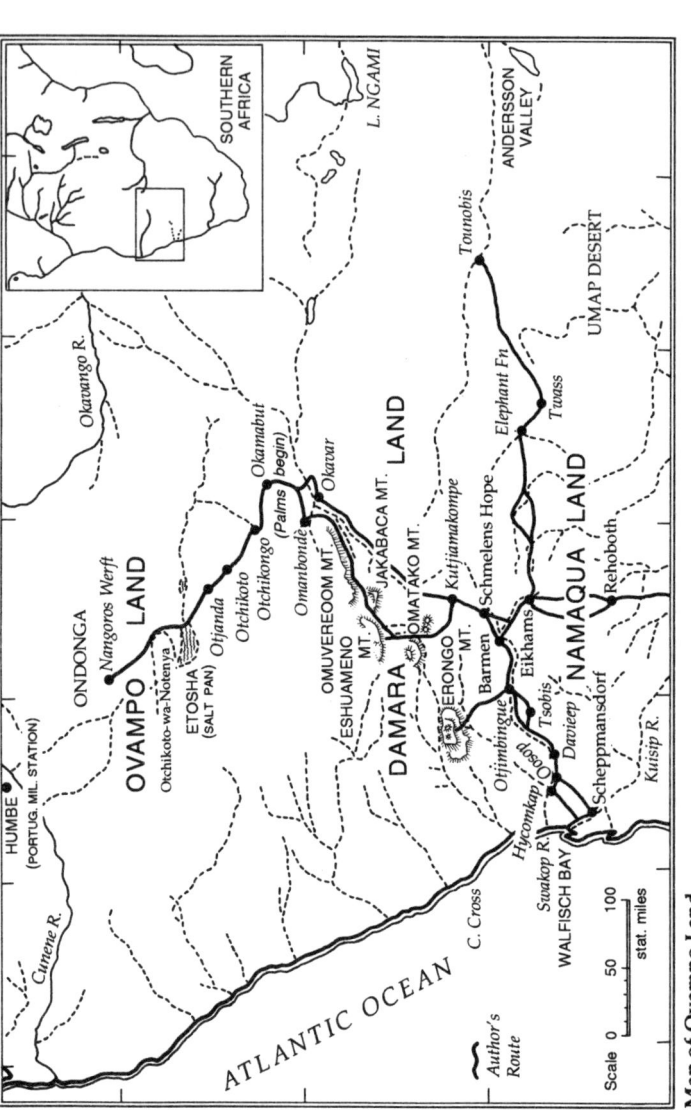

Map of Ovampo Land

(*The dotted river beds indicate temporary torrents occurring some years only during the height of the season. At other times they are dry and sandy.*)

Source: F. Galton, *Narrative of an Explorer in Tropical South Africa* (London: Ward, Lock, 1889). First published under the title *Tropical South Africa* (London: John Murray, 1853).

Figure 3.1 Map of Ovampo Land showing Galton's routes in south-west Africa, 1850–2

for oxen. They keep them to look at as we keep fallow deer',
(Galton, 1889, p. 84). The Ovampo were agriculturists, living in
'charming corn country . . . a land of Goshen'; they were 'decidedly
nice-looking . . . a merry set and all of them dressed, or rather
ornamented very tidily', and 'The ladies were buxom lasses, having
all the appearance of being good drudges' (Galton, 1889, p. 130).

The long slow march across Namibia was, however, more than just
a gentleman traveller's safari. As Galton was to write many years
later: 'It filled my thoughts at the time with enlarged ideas and new
interests, and it has left an enduring mark on all my after life'
(Galton, 1889, p. 193). There was for instance an early (if rather
bizarre) exercise in 'anthropometry', as Galton was to name one of
his favourite ploys – measuring the human frame. Employing his
sextant (from a modest distance) he was able to calculate, by means
of trigonometry and logarithms, the impressive posterior dimensions
of a Hottentot woman. Seeds were also planted of a line of thought to
be elaborated some thirty years later in *Inquiries into Human Faculty*
under the heading 'Gregarious and slavish instincts'; the origin of
these instincts in man he sought to trace in 'the world of brutes'. An
analogy was to be found, he maintained, in the herd instinct he had
observed in his oxen, a habit which was very like the 'slavish apti-
tudes from which the leaders of men are exempt but which are
characteristic elements in the disposition of ordinary persons'. A
traveller with oxen, he suggested, 'is in the position of a host to a
company of bashful gentlemen when he is trying to get them to move
from the drawing room to the dining table and no one will go first'
(Galton, 1911, p. 47ff). But just as a leader will emerge in human
affairs, so will a good 'fore-ox'.

Galton was back in England early in 1852, leaving Andersson to
make his own way to Lake Ngami and in fact to spend the rest of his
life in southern Africa. He was received with acclaim by the RGS,
awarded the Founder's Medal and elected a member of the council in
1854. He was to serve as an honorary secretary from 1857 to 1863.
One of his first tasks, and a continuing one, was to try and supply to
intending travellers at least some of the advice he had found so
woefully lacking when he set out himself. He was closely concerned
with the production of the Society's *Hints to Travellers* (1854; Galton,
1854). First published in 1854, *Hints* was regularly revised and re-
issued for over a century until its simple instructions were overtaken
by the increasing complexity of modern methods and techniques.
Galton's own experience was brought into play with regard to both

organisation and practice. Notably he advised that 'The greatest forbearance and discretion are strongly recommended in all inter-course with the natives. Never allow an imaginary insult to provoke retaliation which may lead to bloodshed. It must be borne in mind their's is the right of soil – we are the aggressors' (RGS *Hints*, 1854).

Parallel with *Hints*, Galton produced his own delightful *Art of Travel: Shifts and Contrivances available in Wild Countries*, which ran into eight editions between 1855 and 1893. Although *Art of Travel* incorporates information gleaned from explorers all over the world, it is Galton's voice which catches the ear. In the dry wastes of Namibia he had devised 'shifts and contrivances' not only in matters of feeding and health, party discipline and clothing, but also of measuring distances and calculating time. 'A man should ascertain his height; height of his eye above ground; ditto when kneeling . . . the span from ball of thumb to tip of one of his fingers; the length of the foot . . .' (Galton, 1872, p. 35). He composed tables for 'the easy determination of the rate of travel, based on the length of pace of strides of men, horses, camels, etc.' (Galton 1872, p. 34). When facing 'the rush of an enraged animal' (which he declares is 'far more easily avoided than is usually supposed') the traveller should bear in mind 'the rate of an animal's gallop', calculated on p. 35. (Galton, 1872, p. 253).

Art of Travel contains instructions for making a sundial, for impro-vising a pendulum by which to make longitudinal observations and for constructing an hourglass (Galton, 1872, pp. 42–3). Bell tents are hopeless – you can't stand upright in them – and he describes a design which allows headroom (Galton, 1872, p. 155). Don't imagine you can take medicines for every possible ailment – powerful emetics and purgatives, eye-washes, quinine for 'ague' is all you want – otherwise improvise. A charge of gunpowder in warm water makes an effective emetic and a scorpion sting can be treated with the oil scraped from the bowl of a pipe (Galton, 1872, pp. 14, 15, 21).

He has sensible advice on the relationship of masters and men; 'Great allowance should be made for the reluctant co-operation of servants, they have infinitely less interest in the success of the expedi-tion than their leaders' (Galton, 1872, p. 5). 'Encourage the men to bring their womenfolk – it adds life to the party' and 'when the women take on the cooking they can live cheaply on what they lick off their fingers' (Galton, 1872, pp. 7–8). And 'Entrust surveying instru-

ments and fragile articles to some respectable old savage, whose
infirmities compel him to walk steadily. He will be delighted at the
prospect of picking up a living by such easy service' (Galton, 1872,
p. 24).

By the time the eighth edition appeared a mass of travel information had been assembled from every continent and arranged by
Galton in the neatest possible way. Arctic explorers are quoted on
the making of pemmican; the African Mansfield Parkyns has a remedy for snake-bite; Charles Darwin offers advice on the manage-
ment of mules; Speke finds a tape run through the waist of his
trousers more reliable than a belt. When visiting Spain with a party of
astronomers to view an eclipse of the sun in 1860, Galton noted with
approval the snug sleeping bags used by 'the French douaniers who
watch the mountain passes of the Pyrenean frontiers'. (Galton, 1872,
p. 150). *Hints* and the *Art* are now part of exploration history;
travellers no longer look for advice on how to make friends with an
elephant (offer her a banana before mounting); how to stop a don-
key's braying (tie a stone to its tail); how to improvise underwater
spectacles; how to swim a horse across a river.

Although his great contribution to scientific thought – the 'legacy'
we celebrate today – does not seem to have been directly inspired by
what he saw in Africa, much else was. As a member of the RGS
Council he did much to upgrade geography as a school and university
subject in a campaign which was to result in the establishment of
university schools of geography. To a large extent Galton was acti-
vated by recollections of his own inadequate education when he
wrote: 'Owing to the unhappy state of education that has hitherto
prevailed by which boys acquire a very imperfect knowledge of the
structure of two dead languages, and none at all of the structure of
the living world, most persons preparing to travel are overwhelmed
with the consciousness of their incapacity to observe with intelligence
the country they are about to visit' (Galton, 1872, pp. 2–3). It was his
idea to institute the award of medals for geographical work by boys in
public schools, a scheme which ran from 1869 to 1884 when it petered
out for lack of support by the schools. This was not altogether to
Galton's regret: 'Geography' he concluded, was an unsatisfactory
subject on which to set examinations: 'You cannot set problems in it',
he wrote to H. R. Mill, RGS Librarian, 'Given the data you cannot
draw references of any high degree of subtlety' (RGS Archive).

Probably Galton's most effective contribution to geographical research was in the realm of meteorology, a subject dealt with in Chapter 4 of this book. His interest in the subject was aroused during his African journey by the practical effects of the weather on the traveller, and he absorbed eagerly the observations of Speke and Grant on Lake Victoria in 1863. For many years he was a member of the Meteorological Council, and he was a pioneer of the mapping of weather, inventing the term 'anti-cyclone'. The first weather map to be published in *The Times* (on 1 April 1874) was by Galton. There was indeed nothing he was not prepared to map, his most eccentric effort being his 'Beauty Map of the British Isles', showing the prettiest girls to be in London, the plainest in Aberdeen – perhaps it is just as well it was never published.

Galton was an enthusiastic participant in the affairs of the British Association for the Advancement of Science, arranging his summer engagements so as to attend its yearly meetings. From 1863 to 1867 he was the Association's general secretary, and he held the important position of president of the geographical section at the Brighton meeting in 1872. This involved him in the heated controversy which surrounded H. M. Stanley, whose address to the meeting on 'How I found Livingstone' was challenged by Galton as offering no useful geographical information.

The travels of Galton's later years were mild affairs compared with his African adventure. In 1853 he married Louisa Butler, whose poor health seems to have encouraged in her husband a tendency to hypochondria. Together they visited spas in Europe each year, until Louisa's death in 1897 in France. Afterwards the 'migratory bird' of his youth seems to have fluttered its wings again and, accompanied by a favourite niece or nephew, he became more enterprising. One year in Spain he attended a bullfight, another journey took him to Greece where he tried out his voice in the classical theatre of Epidaurus, and in 1899 he and his niece Eva Biggs visited the archaeologists Sir Flinders and Lady Petrie in Egypt. He was as full of 'shifts and contrivances' as ever. In her 'Recollections of Francis Galton' (included in Karl Pearson's *Life*) his niece Millicent Lethbridge describes a typical episode during a railway journey in France. 'The heat was terrific and I felt utterly exhausted, but seeing him perfectly brisk and full of energy in spite of his 82 years, dared not . . . confess to my miserable condition. . . . Then with perfect simplicity and disregard of appearances . . . he proceeded to twist his newspaper into the shape of a wash-hand basin, produced an infinitesimally small piece of soap, and poured some water out of a medicine bottle and we performed our ablutions' (Pearson, 1930).

He died in 1911, increasingly feeble in body but mentally alert to the end, the last of the great Victorian polymaths.

References

Galton, F. (1854) 'List of astronomical Instruments, etc.', in 'Hints to travellers', *Journal of the Royal Geographical Society*, vol. 24, pp. 1–13.

Galton, F. (1872) *Art of Travel; or, Shifts and Contrivances Available in Wild Countries*. 5th edn (London: John Murray); first ed. 1855; eighth ed. 1893.

Galton, F. (1889) *Narrative of an Explorer in Tropical South Africa* (London: Ward, Lock and Co.) Second edition of *Tropical South Africa* (1853) (London: John Murray).

Galton, F. (1908) *Memories of my Life* (London: Methuen).

Galton, F. (1911) *Inquiries into Human Faculty and its Development*, 3rd edn, 2nd printing (London: Dent).

Pearson, Karl (1930) *The Life, Letters and Labours of Francis Galton*, Vol. III B, pp. 446–9 (Cambridge at the University Press).

Royal Geographical Society (1854) 'Hints to travellers', *Journal of the Royal Geographical Society*, Vol. 24, pp. 1–13.

Royal Geographical Society: correspondence files, 3 March 1896. University College London Library: Galton Papers 92–103.

4 *Meteorographica* and Weather

Sir Crispin Tickell

For me the name of Galton produces family as well as scientific echoes. Francis Galton was a friend and ally of my great great grandfather, and was one of that heroic group which took on – and then transformed – the intellectual establishment of his day. Recently I looked at my copy of the *Record of Family Faculties*, published in 1884. It has been faithfully filled in for four or five generations: it is good to know the colour of my great great aunt's eyes, whether my cousins were choleric, bilious or sweet tempered, which grandmothers died of what, and who was bright or not so bright in the bell-shaped curves of inheritance. What a pity we do not fill in such forms today!

Galton's long-standing interest in meteorology came through an interest in maps, in turn engendered by his long and generally fruitful association with the Royal Geographical Society. Like many of us he enjoyed making maps; but unlike nearly all of us he went so far as to invent a device to make an ordinary map stereoscopic. Weather maps are of course special. In Galton's day there was no accepted or comprehensive system of symbols for such factors as temperature, wind speed or barometric pressure. In retrospect it seems surprising how little had been achieved in a country so dependent on sailing the seven seas.

When Galton turned his ingenious mind to weather maps around 1860, the Meteorological Office was only five years old. Its principal animator was Vice Admiral Robert FitzRoy (1805–1865), who was the first in this country to attempt weather maps and, on the basis of observations from a few weather stations and lighthouses, to predict storms and issue warnings. FitzRoy's *cone* is still the standard symbol for a gale warning. His work may have been sketchy and amateur, but he was a formidable man in his day, now best remembered as captain of *The Beagle* during its famous voyage between 1831 and 1836. It was he who chose Charles Darwin to be his companion and naturalist. Charles Darwin was Galton's first cousin, and FitzRoy, with his charm, bad temper, brilliance and literalist theology (he resisted evolution with passion), was well known to them both.

I do not know at what point Galton decided to build on FitzRoy's work and attempt both a new symbolic annotation of weather patterns, and the collation of information for this purpose. In 1863 Galton wrote that scientific study of the weather on a global scale was an impossibility for lack of data, and that meteorologists were strangely behindhand in bringing together what there was. There were, he said, more than three hundred skilled observers scattered over Britain and the Continent who transmitted observations to meteorological societies or government institutions, and about the same number of lighthouse keepers who transmitted good returns of at least wind and cloud to one or other of the three Lighthouse Boards of Britain. But no means existed of putting together this mass of information, which came in different forms and measurements. Thus in Britain *fahrenheit* was used for temperature and *feet* for distance; in Germany *reaumur*, in Paris *feet*, and in the rest of France, Belgium, Holland and Switzerland *centigrade* and *meters*.

With characteristic boldness Galton decided to put things right as soon as possible. As a beginning he took a specimen month – December 1861 – and wrote to all the meteorological bodies he could find for the information they could collect, at the rate of three times a day, about weather behaviour in that month. The results were recorded in his own system of symbols in a volume entitled *Meteorographica*, which he published at his own expense in 1863.

It is clear from *Meteorographica* that Galton received a patchy response to his appeal for information. Solid help came from Holland, Belgium, Austria and Germany, a little help came from France and Italy, but none came from Sweden and Denmark. Nevertheless he had enough to go on, and he was helped by the character of European weather in December 1861. The first half of the month was classic cyclonic low pressure weather, and the second half was equally classic anti-cyclonic high pressure weather. Galton had already coined the term *anti-cyclonic*. He was among the first to recognise the distinction between cyclonic weather, in which winds in the Northern Hemisphere circulate anti-clockwise at surface level and clockwise at the top of the column, and anti-cyclonic weather in which the reverse applies.

The observations of 1861 fully confirmed this important observation. He also made two other significant points. First he noticed the simultaneity of wind changes over the vast area under observation. Nevertheless they did not, he wrote, 'move with regularity, ridge behind ridge like waves of the sea, but they are ever changing their contours and other sections'. Secondly he noticed the importance of the Alps, which like other mountain systems broke up and dispersed wind patterns. But he admitted that all his conclusions must remain tentative so long as there were such enormous gaps, particularly over the Atlantic, in available information.

The publication of *Meteorographica* was not only helpfully positive in that it showed what could be done, even with scarce materials, but also helpfully negative in showing how little was known. Considering the rapid advances in science elsewhere, it is perhaps surprising that meteorology remained more descriptive than scientific for almost another half century. Galton himself was asked, following the suicide of Admiral FitzRoy in 1865, to sit on a new Board of Trade committee to look into the possibility of creating a national weather service. This led to the creation of the Meteorological Committee, later to be called the Meteorological Council. His symbols became widely accepted, and his first weather map was published in *The Times* on 1 April 1875.

Galton next turned his ingenuity to better interpretation of the wind charts already available through the Meteorological Office. From these he extracted 'passage charts' in 1866 to help sailors follow a particular course. This led in turn to the creation of 'passenger maps'. The fifth edition of *Hints to Travellers* (1883) had as its frontispiece a map with individual zones indicating the time required to reach all parts of the world from London. Galton constructed this map in 1881 from the timetables of the railway and steamship companies. But the replacement of sail by steam took the heart out of Galton's work in this direction, and this map was his last contribution to geography.

As elsewhere in Galton's work, he started something big but somehow failed to follow it through. His ingenuity in thinking up systems of observation somehow got in the way of his ability to draw general conclusions from the information thus obtained. Looking back from the vantage point of 1991, three points stand out. First we continue to lack information about weather in many parts of the world, notably the southern oceans. At the Second World Climate Conference in Geneva in October–November 1990, it was noted in the scientific declaration that data derived from surface observations were inadequate, and were if anything deteriorating. An appeal followed for a new and ampler system of global observations to improve forecasting. Information from satellites and computers, much beloved of governments, is rarely enough.

We can now understand even better than Galton the interconnectedness of weather worldwide. Perhaps the best example is the phenomenon of *El Niño* off the west coast of South America, whereby every five to seven years or so the winds diminish in strength and fail to blow away the warm surface water from the top of the Humboldt current, thereby causing changes in the behaviour of the world weather machine which can cause floods in California, drought in Northern Australia and even possibly changes in wind and weather patterns across the Sahel.

We also have a better idea than Galton of the precariousness of world weather. A small and haphazard change in one place can have enormous results in another. The flap of a butterfly's wings in Manhattan can eventually lead to a downpour in Regents Park. Indeed the pulse of the ice ages affects, and is substantially affected by, changes in ocean currents. The Gulf Stream, which keeps Western Europe so green and warm for its latitude today, is dependent on the transfer of heat on a long relay belt from the Indian Ocean; a break in that belt could have alarming consequences for us all.

I have sometimes wondered whether Galton ever lifted his eyes from improvements in the description of weather on the basis of thrice daily observations, to deeper and longer changes in climate over hundreds, thousands or millions of years. He was presumably aware of the discoveries by Agassiz and others of past ice ages, with all their implications for the age of the earth, and must surely have tuned in to the energetic debate on that subject in the 1840s and 1850s. He was also a strong supporter of Darwin and Huxley in the battles of the 1860s over the origin of species and the stretching of time and change which evolution involved.

If Galton were alive today I suspect that he would have been primarily interested in the methodology upon which current knowledge of weather and climate is based. Once he had understood or invented a methodology his interest tended to wane. So in 1991 his focus of interest might have been in how to solve the continuing conundrums and invent or perfect methodologies for the purpose.

The establishment of correct chronology is of course critical. Looking back and looking forward require very different techniques. For the past there are two kinds of evidence. First there is the long-term scientific evidence. During the last thirty years many new techniques have been developed – from measurement of carbon 14 to potassium argon – but three are of particular importance for changing weather and climate: measurement of oxygen isotopes in both ice cores and foraminifera for the distant past; pollen analysis for variations in trees and plants in the medium past; and dendrochronology for seasonal variations in the recent past.

Then there is the short-term historical evidence: here also we have learned a surprising amount, some directly and some by inference from historical records, which provide fragments from which understanding of climate change can be derived: the relative advance and retreat of Alpine glaciers; dates in Kyoto for the annual festivals of cherry blossom; the timing of the annual opening and shutting of Baltic ports; the dates of the vendage and harvests in different parts of Europe; the fluctuations of grain prices, indicating supply and demand at particular times of year; the incidence of prayers for rain in Spain; and of course direct accounts of meteorological conditions (as in the diaries of Pepys, Evelyn and others from the seventeenth century onwards). Such historical evidence is of value, but given the complexity and random quality of local, even regional, climatic conditions, and the number of other factors operating, it must be used with caution. Descriptions of 'good' or 'bad' weather are especially misleading: 'good' or 'bad' for whom?

Nevertheless when such evidence is put together and combined with new knowledge of geology and astronomy, a new picture of the past emerges. The Pleistocene ice ages have so far lasted around 2.5 million years. During the last 850 000 years there have been roughly nine advances of the ice (to cover much of North America and Western Europe) and nine relatively brief intermissions or interglacial periods, each lasting between 10 000 and 15 000 years. Our present one began about 14 000 years ago and reached its peak around 6000 years ago.

The reasons for ice ages are not still fully understood, but the key factors were probably the particular configuration of land and sea following movement of tectonic plates; the rise and fall of mountain ranges (Galton spotted the importance of the Alps in deflecting winds); and variations in the earth's orbital relationship with the sun known as the Milankovitch effect: in other words the combination of eccentricity of orbit (100 000 years), tilt (41 000 years) and wobble (21 000 years). Of special importance is the effect of summer radiation on glacier formation in the northern hemisphere. Two recent conclusions are that there is a 100 000 year cycle with a multiplier effect on ocean and ice feedbacks; and that interglacial periods begin with high obliquity of tilt and perihelion in June (in other words when the earth is nearest to the sun).

Let us now look at what has happened in the last 14 000 years, which included a fascinating 800 years of recession to icy conditions some 10 000 years ago. We can distinguish five main periods:

- The hypsithermal, roughly 6000 years ago. It was then about 1.5°C warmer than now with some asymmetry caused by perpetuating ice conditions: in general there was more precipitation, and sea levels were probably higher (perhaps two meters on average).
- Cooling followed, causing a modest glacial advance and more aridity, from 4800 years ago to around 2000 years ago.
- Warmer conditions came again, rising to a peak in certain areas between 900 and 1300 AD: again the Arctic ice retreated and forestation followed: the temperature was perhaps 1°C above present;
- Then came the so called little ice age, beginning in the fourteenth century and continuing, with fluctuations, until the early nineteenth century, with temperature perhaps 1–2°C less than now.
- Warming resumed between 1880 and 1940, to be followed by a levelling off until 1970. Since then there has been a marked warming, with 1990 the warmest year since accurate records began.

Throughout this warm patch in the history of the earth – a blink in geological time – there have been immense regional and local variations, which make description of anything but broad trends hazardous: for example warming in the Antarctic and cooling in the Arctic have taken place simultaneously over the last forty years. Knowledge of events in the southern hemisphere is anyway no more than sketchy.

For the future there are more than enough uncertainties to delight the heart of Galton or anyone like him. In the longer term, and if the Milankovich effect is correctly understood, we shall move back into an ice age, perhaps markedly in 3–4000 years time, with a glacial maximum in 20 000 years time. In the meantime, with the prospect of artificially-induced global warming, things look as if they are going the other way. Indeed they might go so far the other way that natural rhythms might be upset.

Our main instrument for prediction is computer modelling. At present weather forecasting has improved, but not beyond a few days ahead. The complexities of weather will probably continue to defy us.

Longer term trends of climate are easier to predict, and most existing models suggest similar results. These were put together in the report of the Inter Governmental Panel on Climate Change published in September 1990.

These conclusions are being updated, but in the meantime the ones that have attracted most attention are based on the not unreasonable assumption that in spite of all warnings we shall continue to pump greenhouse gases into the atmosphere in increasing quantities. In such a case there is likely to be a rise of the average global mean temperature by about 0.3°C a decade (between 0.2°C and 0.5°C), leading to rise of 1°C by the year 2025 and 3°C by the end of the century. This may seem small, but we should bear in mind that averages cover sharp differences at different latitudes with warming increasing towards the poles (at the coldest time of the last glaciation the average global temperature was only about 4°C or 6°C less than it is now).

There would anyway be marked regional differences. Land areas would be affected more than oceans. Although it is not possible to be precise, southern Europe and North America might have less summer precipitation and lower soil moisture, and the Indian subcontinent a lot more. The world would be generally wetter with less snow cover and ice. There would be some redistribution of weather patterns with drastic local effects. At the same time there would be a sea level rise of around 6 cm a decade, leading to rise of around 20 cm by the year 2030, and around 65 cm by the end of the next century. Throughout we must take good account of the long lag time between cause and effect due to the stabilising and smoothing effect of the oceans.

There are of course many uncertainties, but none, singly or together, affects the main predictions. Among the uncertainties are variations in solar radiation; the role of clouds and the hydrological cycle; the behaviour of oceans as a thermostat; the nature of oceanic/ atmospheric exchanges; the mechanism of the carbon cycle (at present almost half the extra carbon dioxide we are pumping into the atmosphere is unaccounted for); and finally the behaviour of polar ice sheets and sea ice.

Things could prove better or worse in the event. It is easy to be optimistic, and most people probably feel in their bones that something will happen to stave off disaster. But disasters, short or long term, have happened in the past. They may certainly happen again.

Galton would have relished some of these problems. What a pity we no longer have his ingenious mind to find new ways of looking at them! All studies of weather and climate suffer from an irremediable human defect. We are very small, and the sky is very big. Thus the smallest perturbations in the behaviour of the gases that make up the atmosphere have enormous effects on the creatures which scurry about below. As Galton well said: 'No ordinary liquid can give an idea of that combination of mobility and thinness that characterizes the aerial strata, unless it be a film of oil lying on a pool of water'. We are among the microbes under the oil.

References

Galton, F. (1863) *Meteorographica, or Methods of Mapping the Weather* (London and Cambridge: Macmillan).
Galton, F. (1881) 'On the construction of isochronic passage charts', *British Association Report*, vol. 51, pp. 740–1. Also, *Proceedings of the Royal Geographical Society*, vol. 3, pp. 657–8.
Galton, F. (1884) *Record of Family Faculties* (London: Macmillan).
Royal Geographical Society (1883) *Hints to Travellers*. 5th edn.

5 Intelligence and Hereditary Genius

H. J. Eysenck

Ever since Galton raised the question of 'hereditary genius' (Galton, 1869), we have been plagued by the problem of distinguishing between high intelligence and the creativity that is the prerogative of genius. Terman (1925) entitled his study of children with IQs above 140 'Genetic Studies of Genius', ensuring the identification of IQ and genius, but while historical geniuses have usually been rated high in IQ (Cox, 1926), the reverse is certainly not true: there are many highly intelligent people with IQs above 180 whom nobody would call geniuses. Empirical studies of creativity, or ratings of creativity, have usually shown only slight correlation with IQ, suggesting independence between very high IQ and creativity (Prentky, 1980). IQ may be a necessary, but is certainly not a sufficient condition of 'genius'; creativity clearly implies much more than intelligence.

One important additional variable in the cognitive field is constituted by the group of special abilities (Eysenck, 1979), such as verbal ability, numerical ability, visuo-spatial ability and so on. These are independent of g (general cognitive ability), and of each other; thus the variance of a typical test of verbal ability would be written:

$$\sigma_t^2 = \sigma_g^2 + \sigma_v^2 + \sigma_e^2$$

where t stands for the test used, g for general ability, v for verbal ability and e for measurement error. The special abilities of Mozart, Newton, Shakespeare and Titian probably differentiated them much more clearly than would differences in g; hence the importance of such variables. Nevertheless there are many individuals who apparently rate high on tests of these variables but who nevertheless only contribute works of minor distinction; genius must imply more than this! (Andreani and Orin, 1972.)

Psychopathology or mental illness (usually manic-depressive illness or schizophrenia) has often been suggested as a correlate, or even a causal condition of genius: Lange-Eichbaum (1956), Prentky (1980),

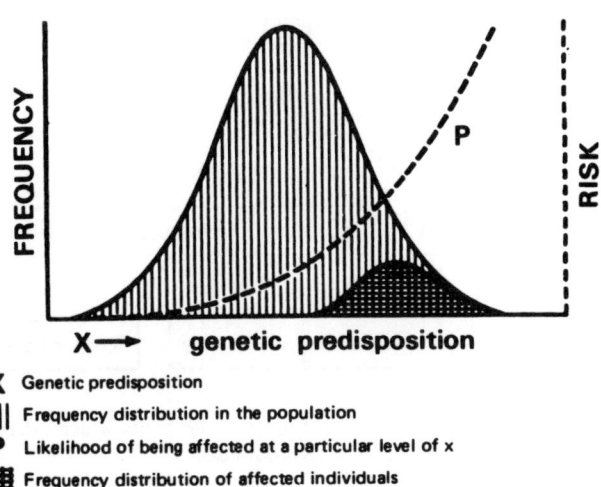

X   Genetic predisposition
|||| Frequency distribution in the population
P   Likelihood of being affected at a particular level of x
▦   Frequency distribution of affected individuals

Figure 5.1   The nature of genetic predisposition to schizophrenia, along the continuum defining the concept of 'psychoticism'

Barron (1963), and Karlsson (1978) have given detailed discussions of the theory that 'great wits are sure to madness near alli'd, and thin partitions do their bounds divide', as Dryden so memorably put it. There is a germ truth in the saying, but the notion that the typical genius is mentally ill is certainly a gross exaggeration. It does contain the germ of a valuable hypothesis, however, namely that creativity is more related to non-cognitive personality traits, such as introversion and a strong ego, rather than to purely cognitive abilities (Taylor and Barron, 1963; Cattell and Butcher, 1968).

The theory put forward here was originally suggested by Eysenck and Eysenck (1976) in the context of the derivation of the major dimensions of personality. In addition to the widely recognised dimensions of extraversion–introversion (E) and neuroticism–stability (N), we advanced the claims of a third dimension, entitled psychoticism (P), based on the view that there is a continuum underlying 'psychotic' behaviour, ranging from one extreme (no predisposition) to the other (high predisposition). Figure 5.1 indicates what is meant by 'predisposition'. Individuals inherit biological structures which, in interaction with the environment, make them likely to show certain behaviours; these constitute the trait, in this case, of 'psychoticism'. The greater the predisposition, the greater the probability of

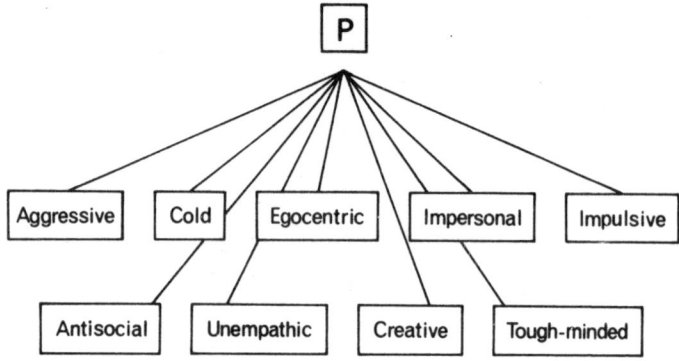

Figure 5.2   Diagrammatic representation of the traits which correlate to
            define the concept of psychoticism

suffering an actual psychotic illness; P indicates this probability,
rapidly increasing as the predisposition increases. The frequency of
different degrees of genetic predisposition is indicated by the large
Gaussian curve; the frequency of actual psychosis by the small,
cross-hatched curve towards the right of the diagram.

What are the traits which, correlate to define psychoticism? Figure
5.2 shows some of these; the literature concerning the adequacy of
the model has been summarised elsewhere (Claridge, 1981, 1983;
Eysenck and Eysenck, 1985; Zuckerman, 1989), and the genetic
contribution to psychoticism has been discussed by Eaves, Eysenck
and Martin (1989). Worthy of note is the inclusion in the list of traits
contributing to psychoticism of the term 'creative'. Is this justified?

Note that it is not being suggested that a genius is necessarily
psychotic, or all psychotics geniuses. What is being suggested is that
psychotics share a common trait of creativity, which by itself is of
little use, particularly when occurring in a person suffering from
psychosis. However in a person high on psychoticism, but *not* mentally ill, and in addition possessed of high g and high special ability,
this creativity, under favourable circumstances, may result in achievements which would justify us in calling such a person a 'genius', I will
first of all consider some of the evidence supporting such a view, and
then go on to consider the particular trait which psychotics and
genuinely creative people may share. I shall begin by considering a
number of genetic studies.

Heston (1966) studied offspring of schizophrenic mothers raised by
foster-parents, and found that although about half showed psycho-

social disability, the remaining half were notably successful adults, possessing artistic talents and demonstrating imaginative adaptations to life to a degree not found in the control group. Karlsson (1968, 1970) in Iceland found that among relatives of schizophrenics there was a high incidence of individuals of great creative achievement. McNeil (1971) studied the occurrence of mental illness in highly creative adopted children and their biological parents, discovering that the mental illness rates in the adoptees and in their biological parents were positively and significantly related to the creativity level of the adoptees. Findings such as these clearly support speculations, such as those by Hammer and Zubin (1968) and by Jarvik and Chadwick (1973), that there is a common genetic basis for great potential and for psychopathological deviation.

These genetic studies may be supplemented by a series of more direct experimental investigations of the hypothesis. These tests have used measures of creativity or of *divergence*, as opposed to *convergence*; convergence implies a single, predetermined answer (for example 2 4 7 11 16 ?), while divergent tests look for indeterminate answers (for example, how many uses of a brick can you think of?)

Assuming for the moment that the P scale does measure, at least to some extent, the essence of the continuum from normality to psychosis, and assuming for the moment that the hypothesis linking creativity and originality with mental abnormality possesses some virtue, then we should be able to test this hypothesis in a variety of ways. It was first tested, in an unpublished study referred to by Eysenck and Eysenck (1976), by D. W. Kidner.

He administered several of the Wallach and Kogan (1965) tests of originality to male and female students, nurses and teachers, and found significant relationships between originality and creativity on the one hand, and high P scores on the other. He also found that 'acceptance of culture', that is agreement with cultural cores, was negatively correlated with P, and also with creativity and originality.

Other studies more marginally relevant to the hypothesis under investigation are reported in the book by Eysenck and Eysenck (1976), but we will turn now to a more recent study by Woody and Claridge (1977) which is particularly impressive.

The subjects of their study were 100 university students at Oxford, both undergraduate and graduate. The students constituted a wide sampling of the various fields of specialisation at the university. They chose students as their subjects because of evidence that creativity is significantly related to IQ up to about IQ 120, but that it becomes independent of IQ above this level (Canter, 1973). The tests used by them were the EPQ (Eysenck and Eysenck, 1975) and the Wallach–Kogan Creativity Tests, somewhat modified and making up five different tasks (instances, pattern meanings, uses, similarities and line meanings). Each task was evaluated in terms of two related variables: the number of unique responses produced by the subject, and the total number of responses produced by the subject.

The Pearson product moment correlation coefficients between psychoticism and creativity scores for the five tests are as follows. P with number of unique responses scores: instances = 0.32; pattern meanings = 0.37; uses = 0.45; similarities = 0.36; line meanings = 0.38. P with uniqueness scores: 0.61, 0.64, 0.66, 0.60, 0.65. It will be seen that all the correlations are positive and significant, and those with the uniqueness score (which is of course the more relevant of the two) are all between 0.6 and 0.7. These values are quite exceptionally high for correlations between what is supposed to be a cognitive measure and a test of a personality trait, particularly when general intelligence has effectively been partialled out from the correlations through the selection of subjects. There were effectively no significant correlations between E and N on the one hand and creativity on the other. It is interesting to note, however, that the L scores (lie, or dissimulation) of the personality questionnaire, which up to a point are a measure of social conformity, showed throughout *negative* correlations with creativity scores, seven out of ten being statistically significant. L is known to correlate negatively with P (Eysenck and Eysenck, 1976).

Studies not using the P scale have come up with traits of 'creative' persons not dissimilar to those characteristic of the high P scorer. Getzels and Jackson (1962) found that 'divergers' were unconventional and independent of judgment. (See also Torrance, 1962.) Hudson (1966, 1968) also noted the conformity of the converger, and the rebelliousness and failure to 'fit in' of the diverger.

It might be said in criticism of the studies so far reviewed, that they deal with psychological tests of creativity and originality in normal and not very distinguished people, and that what is normally understood by originality and creativity demand something more than that. The objection is a reasonable one, although it should not be taken to infirm the remarkable success achieved by Woody and Claridge's empirical testing of the hypothesis linking P and creativity. The only study of what most lay people would consider genuine creativity has been reported by Götz and Götz (1979a, 1979b). Their work significantly extends that of other investigators who tried to link creativity in the arts with personality (for example Csikszentmihalyi and Getzels, 1973; Barron, 1969; Eysenck, 1972; Eysenck and Castle, 1970; Drevdahl, 1956). Some of these studies are difficult to interpret, but we may note that Eysenck (1972) and Eysenck and Castle (1970) found that art students were significantly more introverted and neurotic than non-art students. Götz and Götz (1973) pointed out in criticism that art students in general may not be particularly creative, but when a group of highly gifted art students were compared with less gifted and ungifted subjects, they found that the highly gifted students also had low scores on extraversion and high scores on neuroticism.

In the study under review Götz and Götz (1979a, 1979b) administered the EPQ to 337 professional artists living in West Germany, of whom 147 male and 110 female artists returned the questionnaire; their mean age was forty-seven years. One outstanding result of this work was that male artists were significantly more introverted and significantly more neurotic than non-artists, while for females there was no difference in either of these dimensions. As the authors suggest, it is perhaps true that in our Western World it is mainly women with average or higher scores on extraversion who have the courage to become artists, while the more introverted and possibly more artistically gifted women do not dare to enter the precarious career of the artist.

We must now turn to scores on psychoticism. Here the results are very clear; male artists have much higher P scores than male non artists, and female artists have much higher P scores than female non-artists. As Götz and Götz point out, these results suggest that certainly many artists may be more tough-minded than non-artists. Some traits mentioned by Eysenck and Eysenck may also be typical for artists, as for instance they are often solitary, troublesome and aggressive, and they like odd and unusual things, all high P traits.

The work of Götz and Götz (1979a, 1979b) thus offers important support for the results of Woody and Claridge, and the other authors cited above, in that this more recent work uses actual artistic achievement as a criterion for the measurement of creativity and originality. In doing so they give credence to the validity of divergent thinking tests as measures of creativity and originality, and the fact that in both the artistic and the non-artistic population studied by other investigators significant corelations are found between psychoticism and creativity and originality very much strengthens the hypothetical link between the personality trait and the behavioural pattern. We may thus be justified in concluding that originality and creativity are the outcome of certain personality traits, rather than being cognitive variables or abilities. This is an important conclusion which is somewhat in contrast with assumptions usually made in this field.

The Götz and Götz study is the only one that actually used the psychoticism scale, but other studies implicated traits in creative people which are clearly part of the P syndrome. Thus the work of the Institute for Personality Assessment and Research (IPAR) at Berkeley, under the direction of MacKinnon (1962), was concerned with creativity in architects, writers and mathematicians. As described by MacKinnon et al. (1961) and Barron (1969), creative people showed traits of individualism and independence, lack of social conformity, unconventionality and lack of suggestibility (Crutchfield, 1962); they were also below par in sociability and self-control. Responses on tests like word association were odd and unusual, almost like those of schizophrenics.

Most important, the creative individuals studied by the IPAR group consistently showed greater psychopathology on the Minnesota Multiphasic Personality Inventory (MMPI) depression, hypochondriasis, hysteria, psychopathy and paranoia scales than did the controls. Lytton (1971) concludes that: 'It is difficult . . . to deny that there is more than a chance association between psychiatric difficulties and creative powers' (p. 63). This psychopathology is countered, however, by greater ego-strength, as also shown on the MMPI scales.

The inclusion of introversion and neuroticism in the creativity field needs a little further discussion. Introversion seems to be implicated for both artists and scientists (Götz and Götz, 1979a, 1979b; Cattell and Drevdahl, 1955; Roe, 1952, 1953; Andreani and Orin, 1972, although perhaps more for scientists than for artists (Hudson, 1966). Neuroticism, however, is clearly more associated with the arts than the sciences (Wankowski, 1973; Eysenck, 1978).

We have so far concentrated on creativity as the trait that links genius and high P scorers. It may be useful to note here that the other traits characterising high P scores may also be useful in allowing the scientist or artist with unusual, original, creative ideas to obtain a hearing in a world geared to reward mediocrity (Salieri rather than Mozart); the genius has great difficulties in making his way in the world, and egocentricity, aggression, tough-mindedness and even antisocial behaviour may at times be required to allow his genius expression and social support.

To turn now to a trait which may be characteristic of high P scorers, and of psychotics (particularly schizophrenics), and which was first suggested by E. Kretschmer, who called it *Spaltungsfähigkeit*, or ability to dissociate ideas, as opposed to rigidity of thinking. In excess this might produce schizophrenic thought disorder, inappropriate affect and the maintenance of delusional ideas at variance with everyday experience. More recently, Cameron (1947) has put forward his theory of 'overinclusion', which bears some relation to Kretschmer's concept. According to this view, schizophrenics are unable to maintain the normal conceptual boundaries, and incorporate into their concepts elements which are merely associated with the concept but are not an essential part of it. Payne (1960) has surveyed the literature and Payne and Hewlett (1960) have provided additional experimental support for the theory, which has received a great deal of confirmation. In many tests schizophrenics have been shown to show a high degree of overinclusion in their thinking. This may be regarded as a fault when grossly exaggerated, but it may be recalled that Kretschmer called it an ability (*Fähigkeit*), suggesting that it might be of use when contrasted with rigid exclusion of non-central parts of a concept.

We now have a concept (overinclusion) which may be a clue to what schizophrenics and geniuses share; can this connection be formalised? It is possible to do this within a modern theory of intelligence which goes back to Furneaux's (1960) model of problem-solving behaviour and Campbell's (1960) blind-variation and selective-retention model of creative thought. It is curious that these rather similar conceptions appeared almost simultaneously, although of course they share a long list of progenitors who in part anticipated their theories.

The essence of the Furneaux model is briefly summarised in his own words as follows: 'The brain structure of any individual, P, includes a set of $_pN$ neural elements which participate in problem-solving activities. It is not necessary at this stage to adopt any particular view as to the nature of these elements, which might be either single neurones or much more complex structures. The solution of a particular problem, h, of difficulty D, involves bringing into association a particular set, $_DN_h$ of these elements, interconnected in some precise order. (The terms "bringing into association" and "interconnected" should not necessarily be interpreted literally after the manner of, say, an electrical circuit. For example, the almost simultaneous firing of two otherwise independent units could constitute one method of bringing them into association, provided some device existed which could detect the simultaneity, while the exact order of firing might represent the mode of interconnexion.) When problem h is first presented single elements are first selected, at random, from the total pool $_pN$ and examined to see whether any one of these, alone, constitutes the required solution. A device must be postulated which carries out this examination – it must bring together the neural representations of the perceptual material embodying the problem, the rules according to which the problem has to be solved, and the particular organization of elements whose validity as a solution has to be examined. It must give rise to some sort of signal, which in the case of an acceptable organization will terminate the search process and will initiate the translation of the accepted neural organization into the activity which specifies the solution in behaviour terms. Alternatively, if the organization under examination proves to be unacceptable a signal must result which will lead to the continuation of the search process. It will be found useful to refer to this hypothetical device under the name of "the comparator"' (p. 185).

This model implies the random association of neural elements which eventually generate a solution adjudged correctly by the comparator. Campbell's model, in a similar way, demands 'blind variation' of mental processes; these are then subjected to a consistent selection process to choose for retention those that exhibit 'adaptive fit', that is lead to the correct solution. Both models thus start with the same fundamental idea, and arrive at similar conclusions. Simonton (1988) has used these models in order to give a coherent account of scientific genius; instead of following his detailed application of these models to supreme achievement, we may instead link this general theory to our previous discussion of psychoticism.

According to the Furneaux–Campbell model we are faced with a cognitive problem, and in order to solve it we run through a series of combinations of existing ideas or neural elements which might be relevant to the problem. Finally a correct solution is thrown up, recognised as such by the comparator, and the search is ended. The theory here suggested relates to the number of fundaments (ideas, neural elements) which take part in (are considered for) the search process. Original and creative solutions are more likely to be forthcoming, other things being equal, if the number of elements included in the search process is large rather than small, contains unusual as well as usual components and is overinclusive rather than rigidly narrow. Psychotics, particularly schizophrenics, provide such an overinclusive set of elements but are unable to use the comparator appropriately. They accept as correct conclusions which in fact are erroneous. Low P scorers start out with a limited set of elements which is insufficient to provide novel, original, creative answers. High P scorers, but who are not in fact psychotic, provide *both* the large number of elements necessary for creative answers, *and* the strict comparator which weeds out inappropriate answers.

This, in brief, is a theory of hereditary genius which has much support in the literature, is eminently testable and does not have recourse to mentalistic notions outside the physiological–hormonal–behavioural field of natural science. Genius combines (1) high *g*, genetically determined to a high degree; (2) high specific abilities, also so determined; (3) high P, but combined with high ego strength to avoid actual mental illness; and (4) favourable environmental conditions which enable genius to manifest itself (Amabile, 1983). Undoubtedly there is more than this to genius, and future research will undoubtedly point out the limitations of this model. Nevertheless there is a large amount of evidence supporting some such model, and future research will be more likely to advance knowledge if it builds on these foundations. Genius is a very fuzzy concept, difficult in addition because of its rarity; no wonder science has been so slow to try and understand its roots, or to advance our comprehension of its origins. Nothing more than a beginning has been made in all this, but it cannot be gainsaid that we have made important advances since Galton wrote his seminal books some 120 years ago.

References

Amabile, T. M. (1983) *The Social Psychology of Creativity* (New York: Springer).
Andreani, O. and S. Orin (1972) *Le radici psicologiche del Talento* (Bologna: Societa Editrice Il Mulino).
Barron, F. (1963) *Creativity of Psychological Health* (London: Van Nostrand).
Barron, F. (1969) *The Creative Person and the Creative Process* (New York: Holt).
Cameron, N. (1947) *The Psychology of Behaviour Disorders* (Boston: Houghton Mifflin).
Campbell, D. T. (1960) 'Blind variation and selective retention in creative thought as in other knowledge processes', *Psychol. Rev.*, vol. 67, pp. 380–400.

Canter, S. (1973) 'Some aspects of cognitive function in twins', in G. S. Claridge, S. Canter and W. I. Hume (eds) *Personality Differences and Biological Variation: A Study of Twins*. (Oxford: Pergamon).

Cattell, R. B. and J. Butcher (1968) *The Prediction of Achievement and Creativity* (New York: Bobbs-Merril).

Cattell, R. B. and J. G. A. Drevdahl (1955) 'Comparison of the personality profile (16 PF) of eminent researchers with that of eminent teachers and administrators and of the general public', *British Journal of Psychology*, vol. 46, pp. 248–61.

Claridge, G. (1981) 'Psychoticism', in R. Lynn (ed.) *Dimensional Personality* (Oxford: Pergamon Press).

Claridge, G. (1983) 'The Eysenck psychoticism scale', in J. N. Butcher and C. D. Spielberger (eds) *Advances in Personality Assessment*, vol. 2, pp. 71–114 (Hillsdale, N.J.: Lawrence Erlbaum).

Cox, C. M. (1926) 'The Early Mental Traits of Three Hundred Geniuses', *Genetic Studies of Genius*, vol. 2 (Stanford University Press).

Crutchfield, R. S. (1962) 'Conformity and creative thinking', in H. E. Gruber, G. Terrell and M. Werkheimer (eds) *Contemporary Approaches to Creative Thinking* (New York: Etherton Press).

Csikszentmihalyi, M. and J. W. Getzels (1973) 'The personality of young artists: an empirical and theoretical exploration', *British Journal of Psychology*, vol. 64, pp. 91–104.

Drevdahl, J. E. (1956) 'Factors of importance for creativity', *Journal of Clinical Psychology*, vol. 12, pp. 21–6.

Eaves, L., H. J. Eysenck and N. Martin (1989) *Genes, Culture and Personality: An Empirical Approach* (New York: Academic Press).

Eysenck, H. J. (1972) 'Personal preferences, aesthetic sensitivity and personality in trained and untrained subjects', *Journal of Personality*, vol. 40, pp. 544–57.

Eysenck, H. J. (1978) 'Personality and learning', in S. Murray-Smith (ed.), *Melbourne Studies in Education* (Melbourne University Press) pp. 134–81.

Eysenck, H. J. (1979) *The Structure and Measurement of Intelligence* (New York: Springer).

Eysenck, H. J. and M. Castle (1970) 'Training in art as a factor in the determination of preference judgements for polygons', *British Journal of Psychology*, vol. 61, pp. 65–81.

Eysenck, H. J. and M. W. Eysenck (1985) *Personality and Individual Differences: A natural science approach* (New York: Plenum Press).

Eysenck, H. J. and S. B. G. Eysenck (1975) *Manual of the Eysenck Personality Questionnaire* (San Diego: Edits).

Eysenck, H. J. and S. B. G. Eysenck (1976) *Psychoticism as a Dimension of Personality* (London: Hodder & Stoughton).

Furneaux, W. D. (1960) 'Intellectual abilities and problem-solving behaviour', in H. J. Eysenck (ed.) *Handbook of Abnormal Psychology* (London: Pitman) pp. 167–92.

Galton, F. (1869) *Hereditary Genius* (London: Macmillan).

Getzels, J. W. and P. W. Jackson (1962) *Creativity and Inteland Intelligence* (New York: Wiley).

Götz, K. O. and K. Götz (1973) 'Introversion–extraversion and neuroticism in gifted and ungifted art students', *Perceptual and Motor Skills*, vol. 36, pp. 675–8.

Götz, K. O. and K. Götz (1979a) 'Personality characteristics of professional artists', *Perceptual and Motor Skills*, vol. 49, pp. 327–34.

Götz, K. O. and K. Götz (1979b) 'Personality characteristics of successful artists', *Perceptual and Motor Skills*, vol. 49, pp. 919–24.

Hammer, M. and J. Zubin (1968) 'Evolution, culture and psychopathology', *Journal of General Psychology*, vol. 78, pp. 154–75.

Heston, I. I. (1966) 'Psychiatric disorders in foster home-reared children of schizophrenic mothers', *British Journal of Psychiatry*, pp. 112, 819–25.

Hudson, I. (1966) *Contrary Imaginations* (London: Methuen).

Hudson, I. (1968) *Frames of Mind. Ability, Perception and Self-Perception in the Arts and Sciences* (London: Methuen).

Jarvik, I. F. and S. B. Chadwick (1973) 'Schizophrenia and survival', in M. Hammer, K. Salzinger and S. Sutton (eds) *Psychopathology* (New York: Wiley).

Karlsson, J. I. (1968) 'Generalogic studies of schizophrenia', in D. Rosenthal and S. S. Key (eds) *The Transmission of Schizophrenia* (Oxford: Pergamon).

Karlsson, J. I. (1970) 'Genetic association of giftedness and creativity with schizophrenia', *Heredity*, vol. 66, pp. 177–82.

Karlsson, J. I. (1978) *Inheritance of Creative Intelligence* (Chicago: Nelson-Hall).

Lange-Eichbaum, W. (1956) *Geie, Irrsin and Ruhm* (Munich: Ernst Reinhardt).

Lytton, H. (1971) *Creativity and Education* (London: Routledge & Kegal Paul).

MacKinnon, D. W. (1962) 'The nature and nurture of creative talent', *American Psychologist*, vol. 17, pp. 484–95.

MacKinnon, D. W. et al. (1961), proceedings of the conference on 'The Creative Person', University of California Alumni Center, Lake Tahoe.

McNeil, T. F. (1971) 'Prebirth and postbirth influence on the relationship between creative ability and recorded mental illness', *Journal of Personality*, pp. 39, 391–406.

Payne, R. W. (1960) 'Cognitive abnormalities', in H. J. Eysenck (ed.) *Handbook of Abnormal Psychology* (London: Pitman).

Payne, R. W. and J. H. G. Hewlett (1960) 'Thought disorder in psychotic patients', in H. J. Eysenck (ed.) *Experiments in Personality* (London: Routledge & Kegan Paul), pp. 3–104.

Prentky, R. A. (1980) *Creativity and Psychopathology* (New York: Praeger).

Roe, A. (1952) 'A psychologist examines sixty-four eminent scientists', *Scientific American*, vol. 187, pp. 21–25.

Roe, A. (1953) 'A psychological study of eminent psychologists and anthropologists and a comparison with biological and physical scientists', *Psychological Monographs*, vol. 67, no. 352.

Simonton, K. (1988) *Scientific Genius* (New York: Cambridge University Press).

Spearman, C. (1927) *The Abilities of Man* (London: Macmillan).

Taylor, L. W. and F. Barron (eds) (1963) *Scientific Creativity: Its Recognition and Development* (New York: John Wiley).

Terman, L. M. (1925) 'Mental and physical traits of a thousand gifted children', *Genetic Studies of Genius (Vol. 1)* (Stanford University Press).
Torrance, E. P. (1962) *Guiding Creative Talent* (Englewood Cliffs, New Jersey: Prentice Hall).
Wallach, M. A. and N. Kogan (1965) *Modes of Thinking in Young Children* (New York: Holt, Rinehart & Winston).
Wankowski, J. A. (1973) *Temperament, Motivation and Academic Achievement* (Birmingham: University of Birmingham Educational Survey and Counselling Unit).
Woody, E. and G. Claridge (1977) 'Psychoticism and thinking', *British Journal of Social and Clinical Psychology*, vol. 16, pp. 241–8.
Zuckerman, M. (1989) 'Personality in the third dimension: A psychological approach', *Personality and Individual Differences*, vol. 10, pp. 391–418.

6 The Galton Lecture for 1991: Francis Galton – Numeracy and Innumeracy in Genetics

J. H. Edwards, FRS

The title of this chapter is perhaps unfortunate for innumeracy may appear to be a defect which could be remedied by application, as in illiteracy. However this is not the usage I intend. I know no suitable word to imply the misuse, rather than the disuse, of the application of numbers to the description of things when the use of words or diagrams – Galton's supreme aptitude – would be more appropriate. The theme I wish to develop is that, firstly, Galton's intellectual energy and restless curiosity were adequately contained within the framework of his mathematical knowledge, and secondly that many misunderstandings of both his work, and of the application of mathematics to biology which he did so much to initiate, can be attributed to this.

By this unnecessary and unusual modesty he did much to make respectable the sterile fruit which so often follows collaboration between the mutually ignorant. Galton was clearly a considerable and intuitive mathematician, although, like Faraday who rarely used formulae, his contribution was largely by diagrams dominated by contours and an extraordinary aptitude to visualise in three dimensions and portray in two.

But for a loss of nerve when faced, at the age of sixty-three, with the many deep mathematical problems relating to variation in pairs of measurements, he might have preceded Gregor Mendel (1822–1884) in describing clearly, in English to a wide audience and readership, those discoveries still hidden through the obscurity of the journal in which Mendel published in German. He might also have protected posterity from confounding similarity in twins and other relatives with racism. These are largely the sterile and bitter fruits of the collaboration of the mutually ignorant. Heritability and allied measures of familiarity, which have advanced agriculture by suggesting which traits will respond most rapidly to selective breeding, have been misappropriated and used as an argument both to assert superiorities of sex or race and to denigrate those who have found that relatives are similar by using numerical measures.

This and much else has been attributed to Galton's use of scientific arguments to preserve those privileges of class, race and sex enjoyed by him. It can, however, largely be explained as a result of misunderstanding arguments based on a field of mathematics which Galton greatly advanced, though did not fully understand and, in his later years, lacked the will to either master and challenge or ignore.

Galton's half-cousin, Charles Darwin (1809–1882), never strayed from the confines of his understanding, which was deep and verbal, aided by a few poor drawings of great power. He said of himself: 'I have no quickness of apprehension or wit . . . my power to follow a long abstract of thought is very limited . . . my memory is extensive but hazy'.

On the front cover of the *Annals of Human Genetics*, founded and funded by Galton and whose title was changed by L. S. Penrose (1898–1972) from the *Annals of Eugenics*, there is only one quotation. It is from Charles Darwin and states 'I have no faith in anything but actual Measurement and the Rule of Three'. The inner cover quotes Francis Galton: 'General impressions are never to be trusted', followed by four substantial sentences. The rule of three is now known as the 'metric system' and it once dominated schoolroom arithmetic with such problems as 'if eight men take five days to dig six ditches how many ditches would . . .'

This is the disjunction in the approaches of Charles Darwin and Francis Galton that I wish to develop, initially in relation to the two sides of their extraordinary family, which dominated science and, indirectly, commerce, instrumentation and literature as well as producing Francis Galton and Charles Darwin.

Galton's pedigree, on which his many writings on hereditary genius – summarised in a book of that name – display a modest silence, is remarkable by any standard (Figure 6.1). He and Darwin shared a common grandfather, Erasmus Darwin (1731–1802), who, although living until seventy, died before the birth of either grandson – twenty years before in Galton's case. Their births were separated by over ninety years, an unusual gap, although slightly less than that between T. H. Huxley (1825–1895), who was three years younger than Galton, and his grandson Sir Andrew Huxley (b. 1917).

Erasmus Darwin came from an established family from central

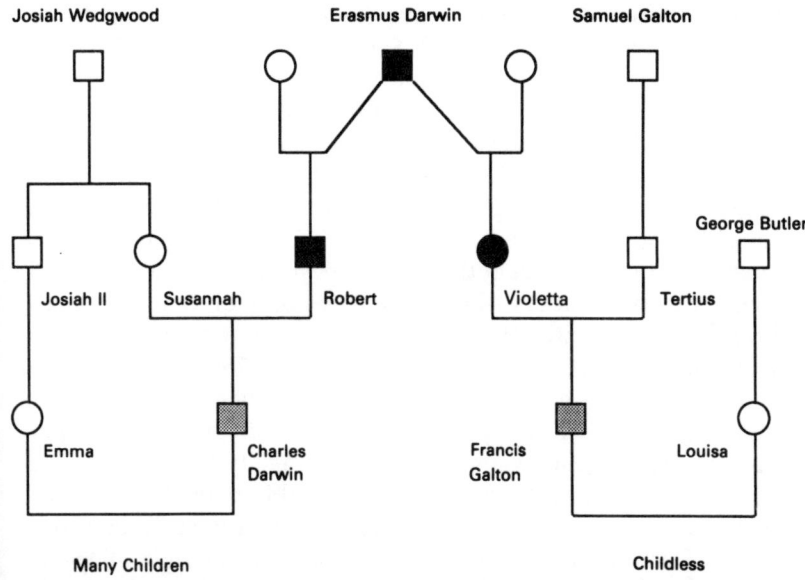

Figure 6.1 Partial pedigree of the Wedgwood, Darwin and Galton
families

England near Nottingham. His grandfather and father lived in what
was effectively the manor house at Elston, a small village famous for
the bloodiest battle of the War of the Roses 250 years before his
birth, and for the fossil of the first ichthyosaur to be discovered in
Britain. This was found in the rector's garden opposite Elston Hall
in 1719, and was reported to William Stukeley (1687–1765) by
Erasmus's father, Robert (1682–1754). The village is perhaps better
known through Godiva, the lady of an earlier manor.

Erasmus Darwin's long and active life, notwithstanding his massive
frame – which required a crescent to be cut from his dining table –
and his aversion to exercise, was dominated by medicine, which he
practiced with sufficient success to refuse an invitation to move to
London as physician to George III. Had he accepted, the king's
ailments might have been better documented and we might have
been saved from the interesting misdiagnosis of porphyria.

Although William Withering (1741–1799) is usually credited with
the discovery of the medicinal use of the foxglove leaf, the source of
digitalis, Darwin has sound claims for precedence. Outside medicine
his many interests included attempts to improve the springing and

stability of carriages. House visits, sometimes as far Bath (he received £40 for visiting the first Samuel Galton (1720–1799), the great-grandfather of Francis Galton), were a major source of income and he always travelled with an extensive library and larder. Two of the carriages he designed survived his estimate of 10 000 miles.

With William Small (1734–1775), who had been tutor to Thomas Jefferson before returning from Virginia to found the Birmingham General Hospital, and Matthew Boulton (1728–1809) he founded the Lunar Society which met at the full moon when travel was easiest, to discuss scientific matters with an emphasis on their application. He also took a major interest in the development of the canal system by his friend and patient Josiah Wedgwood (1730–1795), who was also a Lunarian and later became Charles Darwin's other grandfather. The extensive waterways developed by Wedgwood, initially to import clay and export china whose weight and fragility made road travel uneconomic, were to be even more profitable than the china: the development of canals was largely responsible for the private income Charles Darwin was to acquire at marriage. Thus *Origin of Species* was funded by water: Galton's work, in contrast, was funded by blood (from the proceeds of a gun factory). Charles Darwin appears to have had little contact with Francis Galton, thirteen years his junior, beyond parental visits between Birmingham and Shrewsbury. Their first known letters to each other were very formal.

Erasmus Darwin resolved the problem of windmills needing to be turned into the wind, and requiring mechanisms frequently destroyed in gales, by building a substantial structure with a circular tower and central rotor. This system worked for eleven years until surplanted by the steam power of Boulton and James Watt (1736–1782). He experimented with an artificial larynx, a design reconsidered recently after the failure to achieve acceptable electronic speech. A major poem, 'The Loves of the Plants' (1789), written after translating Linnaeus and often criticised for extensive sections of little merit, hardly survived its parody, 'The Loves of the Triangles'. However it did influence the poetry of Shelley, Wordsworth, Keats and Byron, and Samuel Taylor Coleridge (1772–1834) considered him 'the first literary character in Europe', a view Coleridge later repudiated. Darwin published this work anonymously as he feared a reputation for poetry would compromise his medical practice. The full work, *The Botanic Garden* (1789 and 1791) comprised two volumes totalling 551 quarto pages. *Zoonomia* (2 vols. 1400 quarto pages) discussed evolution with sufficient force to be placed, as Charles Darwin pointed out, on the *Index Expurgatorius*. (E. Darwin, 1794 and 1796) The term 'Darwinising' relating to evolution preceded *Origin of Species* (1859) by more than half a century.

Charles Darwin's other grandfather, Josiah Wedgwood, who left half a million pounds when he died, was the thirteenth child of a cottage-industry potter and later became a friend and patient of Erasmus Darwin. From childhood he was handicapped by a permanently bent knee. This made it difficult for him to manage a potter's wheel, so he preoccupied himself with developing new techniques and designs, as well as surveying the potential routes of canals. He developed the present Wedgwood pottery industry and much of the present Midlands' canal system, eventually achieving the distinction, printed on his card, of being 'Fellow of the Royal Society and Potter to the Queen'. With his family – clothed and horsed as gentry – he was painted by Stubbs. Lancelot Hogben (1895–1975) considered that the demand for lenses created by the need for surveying of sufficient accuracy to ensure that a canal was dug at a constant level, was responsible for the development of lenses adequate for the compound microscope, which created a revolution in biology and medicine (Hogben, 1933). Josiah Wedgwood's industrial empire passed to his second son, Josiah II (1796–1843) who, like Galton's father, moved away from his father's workplace. For a time he lived in Dorset where he was introduced to Coleridge by his brother Tom (1771–1805), who had also been educated at home. Tom is credited with capturing the first photograph using silver salts and strong sunlight: however the plates were unfixed and had to be examined in dim light, fading with each viewing. Tom became the friend and patron of Wordsworth as well as Coleridge, providing the latter with an annuity and sufficient funds to travel in Germany.

While staying with Coleridge at Stowey in Somerset Tom Wedgwood was advised, in a letter from Erasmus Darwin, that his symptoms would benefit from '3/4 of a grain, or a grain, of opium, taken every night for many months . . .'. This advice was possibly fatal to the patient, who died eleven years later aged thirty-four, and to Coleridge's later work, even if it may have spawned 'Kubla Khan' (1797). At a later meeting, within a year of Tom's death, they enjoyed a consignment of hashish sent at Coleridge's request by the president of the Royal Society, Sir Joseph Banks (1743–1820).

The eldest daughter of Josiah I, Susannah (1765–1817), married Erasmus Darwin's son Robert (1766–1848), while the daughter of his grandson, the second Josiah II, was later to marry their son Charles

Darwin. Robert Darwin distinguished himself as a physician, an occupation virtually imposed on him by his father, whose principles he appears to have followed with success but with limited curiosity towards other subjects. It was the Josiah II who was to intercede successfully on behalf of Charles Darwin when Fitzroy invited him as his companion on the *Beagle*.

Galton's paternal grandfather, Samuel John (1753–1832), was a Quaker, a banker and an early industrialist. Although conforming to the dress and language of his sect, retaining the use of 'thee' and 'thou', his factory in Soho, Birmingham was eventually to produce a gun a minute. Because of this the Society of Friends attempted, without success, to ban him from their meetings. He had wide scientific interests and was well known to Josiah Wedgwood and Erasmus Darwin through membership of the Lunar Society. His son Samuel Tertius (1783–1844) maintained and expanded his father's banking interests and married Frances Violetta (1783–1844), Erasmus Darwin's daughter by his second wife. As well as sharing and extending his father's interests he preceded Thomas Young (1773–1829), a fellow Quaker, in advancing, on less firm foundations, the trichromatic theory of colour vision. He slowly dropped his Quaker habits and adopted the church-going of his wife. His son Francis was later to devote much time to fishing, hunting and shooting, the leisured pursuits of the rural upper classes.

Francis Galton, the last of six children after a gap of seven years, was born at 'The Larches', Sparkhill, Birmingham in 1822. This house had been rebuilt on the ruins of the house that had belonged to Joseph Priestley (1733–1804), a Unitarian and Lunarian, which had been ransacked and fired by a crowd following his reputed attendence at a dinner to commemorate the first year of the French Revolution, a fire which 'destroyed the finest laboratory in Europe and warmed the heart of George III'. The ruins were purchased by Withering, whose wine cellar had diverted the rioters from his fine house, now the Edgbaston Golf Club, the first house in England with double glazing. The two larches shown in etchings were probably about thirty years old and were said to be the first of their kind to be introduced to England.

Such were the backgrounds of Darwin and Galton. They were both products of the early industrial revolution, and were largely created by the intellectual commonwealth unrelated to religion or class which arose around those prohibited by religion, or lack of religion, from the universities and discouraged or prevented from living in cathedral cities. Their grandparents were all dissenters or Quakers – Pearson notes that at least eleven, and possibly thirteen, of Francis Galton's sixteen great great grandfathers were Quakers.

Galton's strange childhood, in which he was almost adopted by his younger sisters, the youngest being twelve years his senior, is well described in his autobiography and elsewhere. The bond was peremptorily broken by his being sent, at the age of seven, to a boarding school in France, an act of logic and insensitivity typifying the directness and tactlessness Francis Galton was himself later to show.

His brief apprenticeship to medicine, mainly surgery, in Birmingham was followed by Trinity College, Cambridge, where he read mathematics. This was interrupted by some form of nervous breakdown and was never completed. His father's death at this time left him with sufficient wealth to require no regular employment, while his training provided both a background of biological common sense and a solid basis for his profound mathematical reasoning, which was based, as was that of R. A. Fisher (1890–1962), on the ability to visualise curves, humps and contours. Galton's abilities, like Erasmus Darwin's, also included the ability to visualise related mechanical devices. Unlike Fisher his mathematical imagination was restricted to three dimensions, as we know from his *Inquiries into Human Faculty* (1883).

For ten years he travelled and interested himself in travel, was active in the affairs of the Geographical Society, and spent two years exploring what is now Namibia in South West Africa with a numerous retinue, ox carts and several dozen oxen. He studied the Damar language but, although a deep observer of the behaviour of men and oxen and a meticulous surveyor, showed little interest in the natural history of the region.

After returning at the age of thirty-one he married Louisa Butler (1822–1897), the daughter of George Butler (1774–1853), Dean of Peterborough and previously headmaster of Harrow. Butler was a distinguished classicist, linguist, mathematician and athlete and came from a family well populated by headmasters and churchman: R. A. Butler was his great grandson. This marriage, especially its childlessness, has fascinated many who have found rare explanations of common events. There is some evidence of Louisa's apparently infectious hyponchondria, and the cold and punctilious nature of their letters to and about each other is well documented. As with Charles Darwin's marriage, it appears to have been neither initiated nor sustained by any very strong passions. Childlessness appears to have both distressed and interested Galton deeply: he even produced a graph, of remarkable symmetry, with contour lines showing the probability of childlessness by the age of each partner at marriage.

After settling in London he started his activities in the acquisition of data, developing novel methods for their display and for the measurement of their variability. He invented and produced superficially simple machines capable of drawing ellipses, displaying binomial distribution and acting as simple analogue computers. He was among the first to appreciate the biological significance of variation, as opposed to the prevailing view – espoused by L. A. J. Quetelet (1796–1874) – which regarded variation as the result of inadequate reproduction of an ideal type. His methods of display primarily involved various contours being applied to a wide variety of data, which effectively smoothed the untidiness of the world of discrete numbers. He developed the first simultaneous weather map, acquiring data from many centres at the same time, and proposed various symbols for wind, cloud and so on. His major contribution to meteorology was to propose the model of cyclones and anticyclones as giant vortices, and to name the anticyclone. In 1863 he even attempted the difficult but important mapping of lines of minimal time routes for sailing ships, an advance to be eclipsed by steam but now relevant to airliners.

In 1869 he published *Hereditary Genius*, collecting data from numerous families of distinction and revealing a concentration of both general and specialised talents beyond that which could reasonably be attributed to a privileged environment. Darwin wrote: 'My dear Galton, . . . You have made a convert of an opponent in one sense, for I have always maintained that excepting fools men did not differ much in intellect, only in zeal and hard work . . .'

In his attempt to distinguish the effects of what he termed nature and nurture he appreciated the significance of twins, and was among the first to draw attention to *identical* as being distinct and without intermediate forms, and to describe their origin lucidly and correctly in terms of number of ova. Unfortunately it is in relation to the relative similarities of the twins that Galton's approach is best known and most misapplied. It is in the anecdotal, rather than the statistical, treatment of twins that most has been learned, both from similar intellectual development in identical twins with widely differing background and from dissimilar disorders in those genetically identical.

The similarity of relatives initiated his interest in family measurements; later his aim to improve the next generation by various means of persuasion and exhortation which was to develop into the Eugenic Movement. His studies on the similarity of relatives and other groups, defined by disease or criminality, included a novel use of photography by exposing the same plate for several individuals, each exposure being a fraction of the normal exposure equal to the reciprocal of their number. He had limited success with criminals and disease, but his studies of relatives led members of their family to mistake some composites for individuals who had not been photographed. He also applied subtraction methods, superimposing negatives and positives to define differences.

In a further attempt to unravel the inheritance of ability he enquired into the mechanisms of memory and recall, finding extreme differences in the way individuals remembered events and handled abstractions, the mind's eye varying from a photographic clarity in two or three dimensions in some individuals to an inability even to recall the image of a breakfast table, as with Charles Darwin. This, now best known from his *Inquiries into Human Faculty* (1883), his only book now in print, after the Everyman edition was reprinted by the Eugenics Society in 1951.

Following Darwin's asking his advice on the comparison of measurements relating to self-fertilised and cross-fertilised plants he started experimenting on peas (Darwin recommending sweet-peas as opposed to Mendel who, unknown to either of them, had worked on the edible pea), pursuing the segregation of seven characters in a plant now known to have seven pairs of chromosomes. He persuaded and commissioned a large and well-structured programme of breeding from seed peas of various sizes and collected the resulting seeds, measuring their mean diameter with precision by lining them up by hundreds in grooves. He found that while large peas begat large peas, the parental influence appeared incomplete, the second generation differing from the mean pea size by only a third of the difference by which the parents differed from the parental mean.

This consistent feature, by which like only partially bred like, he termed 'reversion', the pea 'reverting to what may be roughly and perhaps fairly described as the average ancestral type'. The term and concept of 'reversion' to an ancestral type was used widely by Darwin and Wallace (1823–1913). Later he used the words 'regression', and 'regression to mediocrity' – words which convey an unfortunate implication, although in Galton's time mediocre was without its inappropriate modern connotation of being below average. Galton then proceeded to measure families with great thoroughness, and

Table 6.1 Distribution of light-eyed* and other eye colours by their distribution in grand-parents. All the parents whose parents were light-eyed were light-eyed.

Grand Parents	Parents	Children
All light	Both light (all)	174/183
3/4 light	Both light	46/ 53
All	Both light	344/355
2/4 light	1/2 light	30/ 66

* In the category 'light-eyed' Galton included light blue, blue, dark blue, grey and blue-green.

Source: Abstracted from data presented by Galton in Table 19 of *Natural Inheritance* (London: Macmillan, 1889).

also, by inspection and correspondence, obtained substantial data on eye colour (Table 6.1).

In his measurements of fathers and sons and mothers and daughters, as well as persons and adult children, height was standardised by increasing the female height to the male equivalent by adding an inch per foot, a close approximation typical both of Galton's practicality and of his regarding the male as the norm. He produced extensive data on pairs of similar measurements and on the same measurement of pairs of individuals. The data on the pea and on humans were later supplemented by data arising from studies on moths, and with his compulsion to represent everything possible by contours he developed his well-known model, now termed the bivariate normal surface, which was perhaps the most important statistical discovery of the last half of the nineteenth century. It has the elegant feature that whichever way it is sliced vertically the section is a normal curve, parallel slices giving curves of varying size but constant shape, while every horizontal slice yields ellipses varying in size but not in shape.

Galton was fascinated by the beauty and symmetry of this surface, in that, so far as the measurements went, the grown child could be the father to the man without evident change. As vertical slices would clearly have their mode below the main diagonal there would necessarily be regression to the mean. The modes would also form a straight line, which would be the tangent of a vertical or horizontal line to the contour at that point. He built models from cardboard, and had made a substantial machine to scratch plates from which accurate ellipses could be printed. To find an elegant mathematical model representing heredity in peas and man could hardly fail to excite its discoverer, but the analytical solution defeated him.

It was then that misfortune occurred. Forrest (1974) informs us that as Galton had forgotten the handling of conic sections, which he doubtless learned at Cambridge, he went to the library of the British Academy to consult their books, only to meet the president, Sir James Dewar (1842–1923), who advised him not to bother but to consult his nephew, Hamilton Dickson (1849–1931), a mathematician in Cambridge. He took this advice and, on receiving his solution, wrote:

> I may be permitted to say that I never felt such a glow of royalty and respect towards the sovereignty and magnificent sway of mathematical analysis as when his answer reached me, confirming, by purely mathematical reasoning, my various and laborious statistical conclusions with far more minuteness than I had dared to hope. For the original data ran somewhat roughly, and I had to smooth them with tender caution. His calculation *corrected* my observed value of mid-parental regression from 1/3 to 6/17.6, the relation between the major and minor axis of the ellipse was changed 3 per cent, their inclination was changed less than 2 degrees (Galton, 1885, my italics).

Galton, then sixty-three, was so impressed that he considered his observations 'corrected'. When Fisher examined Mendel's observations on qualitative variations in the edible pea he also found the results too good to be true, and had to assume some bias in recording or, worse, some 'corrected' observations. Fascinated by the apparent obedience with which nature followed art Galton resolved the problem of regression by assuming that what was not expressed must be latent, so deriving his theory of ancestral inheritance, whereby every ancestor is proportionally represented, a theory which, by confounding the probability of the origin of what Mendel termed a factor, and the proportion of these factors from any ancestor, introduced a fundamental confusion to both statistics and genetics. This was consolidated by the mathematical identity through which the distribution of a variate and the distribution of the expected mean of a variate on repeated sampling could be expressed and handled.

In 1889, in his *Natural Inheritance* (p. 9), he stated 'So in the process of transmission by inheritance, elements derived from the same ancestor are apt to appear in large groups, just as if they had clung together in the pre-embryonic stage, as perhaps they did'. But such a clear appreciation of a mechanism now known to exist was eclipsed by the mathematics he could not understand – but would have understood sufficiently but for fatal advice to delegate. The verb 'to think' has no plural form. In *Natural Inheritance* the numerous tables on eye colour in families show the fatal power with which the smooth dress of real numbers can conceal the rough bodies of the numerable units which underlie inheritance.

Galton was so fascinated by the success with which height could be tamed by real numbers that he applied them to the integers defining numbers of families with various eye colours in grandparent, parent and child (Table 6.1). If two grandparents have dark eyes and two have light eyes then, according to ancestral theory, it does not matter which two have which, so that families with each parent having one dark-eyed parent and with one parent having two dark-eyed parents can be combined both to save tabulations and to lead to larger numbers in each table. If Galton had split his grandparents by type he could hardly have failed to discover what was later termed Mendelism, and even if Mendel had discovered it twenty years earlier, an independent discovery expressed forcibly in clear English would have lead to Galtonism becoming the established term. We would then have had both Daltonism and Galtonism as descriptions of the atomic nature of chemical and biological processes.

The apparent success of this application of mathematics led to the establishment of a mathematical school of genetics which became increasingly divorced from mechanism, with a preference for studies on organisms easily measured, such as crabs, and an elaboration of descriptive arithmetic involving distributions defined by several moments, each to many decimal places. For example, Karl Pearson (1857–1936) regressed the birthweight of daughters against their father's income in shillings to give 11.381 + 147174R lbs. Sons were predicted to be five pounds heavier if the paternal income was zero (Karn and Pearson, 1922).

An early casualty of this exuberant numeracy was the clear use of language. Pearson assumed that it would be possible to partition the effects of nature and nurture by numerical means, although the full development of an analysis that could allow this had to await the most important paper this country produced on mathematical genetics during the first twenty years of this century. This was submitted by R. A. Fisher to the Royal Society, was rejected by Pearson and Punnett, and was later published in the *Transactions of the Royal*

Society of Edinburgh in 1918 with funding from Leonard Darwin (1850–1943). In 1908 a paper by G. H. Hardy (1877–1947) also had the accolade of rejection, this time by *Nature*, but it was eventually published in *Science* (Hardy, 1908).

The concept that it is meaningful to separate nature and nurture by some sort of proportional representation has been dealt with by many authors, including Hogben in '*Nature and Nurture*', his Withering Lectures of 1933 (Hogben, 1933). Proportional representation, or some such measure as heritability, is of value as a guide to speed of response under selection, but even if it were possible to make estimates of a species with familial environmental similarities, the estimates derived would have little relevance to such issues as the need for education. This impracticality of splitting nature and nurture into proportions was most lucidly discussed by Leonard Darwin, the patron of Fisher and E. B. Ford (1901–1988), in a letter in *Eugenics Review* (L. Darwin, 1913) reprinted at the end of Chapter 7 of this book. This preceded the term 'heritability' and the mathematical framework of the partitioning of variances advanced in Fisher's paper of 1918, largely on data of human height.

The misapplication of heritability arguments to human affairs is still a major cause of public aversion to genetical enquiry, as in the reputed fraud by Cyril Burt (1883–1971). While there is evidence of a casualness of recall with age, especially when his department was evacuated to the Welsh coast during the bombing of London, his measures of similarity between identical twins was consistent with other work, with work on measurements easier to make, such as height, and with his earlier work with collaborators of known integrity. What is disturbing is that these values were irrelevant to the various charges made against his work, implying that the basic nature of regression is widely misunderstood by many who use it. Unfortunately this was not the only major diversion of scientific funds and talent following the development of a mathematical superstructure unsupported by either sound foundations in logic, or a satisfactory way of relating the resulting numbers to the physical mechanisms capable of producing the observations.

The present aim to produce a complete inventory of the human genome base by base is at least partly fuelled by 'new genetics', which, in its original formulation, followed a paper in which the lucidity of the description of the techniques available was confounded by a strategy for linkage analysis that overlooked the low resolving power of the methods proposed. 'New genetics' rapidly came to mean

'DNA methods' and led to several lucid and influential books and a welcome 'new look' to medicine. Unfortunately new genetics led all too quickly into 'new eugenics', in which intrinsically uncertain methods are used to define fetal disease and allow doubtful abortion. The problem is not that such methods are not justified while awaiting a mechanism which allows more exact diagnosis or effective treatment, but that excessive resources devoted to the development of such indirect methods, which result from numerical misunderstandings, are starving basic research into mechanism.

The misapplication of mathematics to genetics has a long history extending from the predictions of astrology to the genome project, and will doubtless continue, and indeed expand, with the development and distribution of inappropriate computer programmes which, however appropriate they were when they were developed, are usually doomed to lag behind technical advances in making observations, so that their half-life as a useful aid is shorter than the half-life of their usage, which usually reaches a maximum after their relevance is waning. The remedy is in the hands of editors and assessors of competence on the basis of reprints, for much of the problem is due to multi-authorship by the mutually ignorant, and the respect that is given to such papers if one of the authors has, or even had, a sound reputation.

The assessment by William Bateson (1861–1926) of Galton's influence is not without justification, for within the field of Bateson's genetical enquiry, a field restricted by his depth of thought and an aversion to non-integral numbers, Galton's influence, largely through Pearson, was not conducive to the development of such studies. He wrote:

> Of the so-called investigations of heredity pursued by extensions of Galton's non-analytical method and promoted by Professor Pearson and the English Biometrical school it is now scarcely necessary to speak. That such work may ultimately contribute to the development of statistical theory cannot be denied, but as applied to the problems of heredity the effort has resulted only in the concealment of that order which it was ostensibly undertaken to reveal (Bateson, 1909).

Hogben, whose deep mathematical insight was, like that of A. S. Eddington (1882–1944), based on numbers he could visualise had similar reservations, and wrote:

Observations upholstered with sufficient algebraic sophistication are *ipso facto* praiseworthy on that account. By the same token, we discount the contribution of the naturalist or of the physician with clinical judgment unless the presentation of the observational record conforms to a ritual whose credentials lie outside the curriculum of his training. More and more he learns to rely on the interpretation by recourse to methods which he understands less and less (Hogben, 1954).

However it is only fitting to end with Galton's words at the close of his best known work, *Inquiries into Human Faculty* (1883), which is largely an anthology of his many biological papers. He wrote, on eugenics, a word subsequently so associated with misapplications which he should hardly be criticised for not forseeing, so that it has done much to impair his reputation. He would have been the first to disown and publicise the abuses to which the living were exposed in the first half of this century by immigration policies and some of the more extreme measures now proposed for the termination of the unborn, which it is hardly inappropriate to term 'new eugenics'.

After a strong plea for accurate and permanent records to be maintained as part of what he envisaged as some sort of national health service, a service now increasingly reluctant to finance the preservation of hospital records for longer than a tenth of a lifetime, he finished his anthology of his major works with:

the new duty which is supposed to be exercised concurrently with, and not in opposition to the old ones upon which the social fabric depends, is an endeavour to further evolution, especially that of the human race (Galton, 1883).

References

Bateson, William (1909) *Mendel's Principles of Heredity* (Cambridge at the University Press) pp. 6–7.
Darwin, Charles (1859) *On the Origin of Species by Means of Natural Selection or the Preservation of Favoured Races in the Struggle for Life* (London: John Murray).
Darwin, Erasmus (1789 and 1791) *The Botanic Garden: a Poem, in Two Parts*: Part I, *The Economy of Vegetation* (London: J. Johnson, 1791); Part II, *The Loves of the Plants* (London: J. Johnson, 1789).

Darwin, Erasmus (1794 and 1796) *Zoonomia; or, The Laws of Organic Life*: Part I (London: J. Johnson, 1794); Parts I–III (London: J. Johnson, 1796) (2nd edn of Part I, corrected) 2 vols.

Darwin, Leonard (1913) 'Heredity and Environment', correspondence in *Eugenics Review*, vol. 5, pp. 153–4.

Fisher, R. A. (1918) 'The correlation between relatives on the supposition of Mendelian Inheritance', *Transactions of the Royal Society of Edinburgh*, vol. 52, pp. 399–433.

Forrest, D. W. (1974) *Francis Galton: The Life and Work of a Victorian Genius* (London: Paul Elek).

Galton, F. (1869) *Hereditary Genius* (London: Macmillan).

Galton, F. (1883) *Inquiries into Human Faculty and its Development* (London: Macmillan); second edition, 1892 (Macmillan); third edition, 1907 (Everyman's Library, London: J. M. Dent & Sons), reprinted in 1911, 1919, 1928, and in 1951 for the Eugenics Society.

Galton, F. (1885) 'Regression towards mediocrity in hereditary stature', *Journal of the Anthropological Institute*, vol. 15, pp. 246–63.

Galton, F. (1889) *Natural Inheritance* (London: Macmillan).

Galton, F. (1908) *Memories of my Life* (London: Methuen).

Hardy, G. H. (1908) 'Mendelian proportions in a mixed population', *Science*, vol. 28, pp. 49–50.

Hogben, L. (1933) *Nature and Nurture; Being the William Withering Memorial Lectures on 'The Methods of Clinical Genetics'* (London: Williams and Norgate).

Hogben, L. (1954) *Statistical Theory* (London: Allen and Unwin).

Karn, M. N. and Karl Pearson (1922) 'Study of the data provided by a baby-clinic in a large manufacturing town', in *Drapers' Company Research Memoirs: Studies in National Deterioration*, vol. X (Cambridge University Press).

Pearson, Karl (1914, 1924 and 1930) *Life, Letters and Labours of Francis Galton*, 3 vols (Cambridge at the University Press).

7 Galton, Karl Pearson and Modern Statistical Theory

A. W. F. Edwards

INTRODUCTION

'To work for that Galtonian renascence has been the writer's main aim in life' wrote Karl Pearson in April 1914, and for us to explore the extent to which Pearson was successful in transmitting and elaborating his Galtonian statistical inheritance it is natural to start with the work from whose preface this quotation is taken, *The Life, Letters and Labours of Francis Galton*, published in three volumes (but four parts) by Cambridge University Press between 1914 and 1930 (Pearson, 1914–30). That renascence was being produced in innumerable branches of science, wrote Pearson, by the ramifications of Galton's methods, and 'will be as epoch-making in the near future as the Darwinian theory of evolution was in biology from 1860 to 1880'. Pearson, having in his preface just taken a swipe at William Bateson and the Mendelian school, added for good measure '. . . and which has encountered and will encounter no less bigoted opposition from both the learned and the lay'. He was not far off the mark, but it was to be Pearson's own statistical work rather than Galton's methods that encountered opposition.

In his Herbert Spencer Lecture delivered in Oxford on 5 June 1907 Galton, looking back over his long life, thought that the publication of his book *Natural Inheritance* in 1889 was the timely catalyst which had launched his biometric methods on their successful path and added: 'The methods were greatly elaborated by Professor Karl Pearson, and applied by him to Biometry' (Galton, 1889b, 1907). A modern commentator, T. M. Porter, agrees: 'The almost simultaneous appearance of Galton's book . . . and his method of correlation in 1889 marks the beginning of the modern period of statistics' (Porter, 1986). C. D. Darlington, in his 1962 introduction to the reprint of Galton's *Hereditary Genius*, called Galton 'the father of biometry' (Galton, 1962).

Yet as Professor S. M. Stigler pointed out in his 1986 R. A. Fisher Memorial Lecture, 'Francis Galton and the unravelling of the Normal world', and repeated in his invaluable book *The History of Statistics* (Stigler, 1986), Pearson was not initially enthusiastic. Pearson was a member of the Men's and Women's Club, which met to discuss improving the relations between the sexes in an atmosphere friendly to both socialism and feminism, and on 11 March 1889 he read to the club a paper about Galton's *Natural Inheritance* whose manuscript is extant. Stigler writes of it: 'Pearson's commentary on Galton was lucid, well-organized, mostly correct, and as perceptive as most, if not all, other commentators. Nevertheless, it was a view by a man wearing blinders', and 'Pearson thus had a first look at Galton's statistical methods but was not able to see their promise. Far from viewing them as techniques for a general class of problems, he remained blinded by restrictions learnt at Cambridge'. Stigler reprints the following extract from the opening pages of Pearson's talk:

> Personally I ought to say that there is, in my own opinion, considerable danger in applying the methods of exact science to problems in descriptive science, whether they be problems of heredity or of political economy; the grace and logical accuracy of the mathematical processes are apt to so fascinate the descriptive scientist that he seeks for sociological hypotheses which fit his mathematical reasoning and this without first ascertaining whether the basis of his hypothesis is as broad as that human life to which the theory is to be applied.

Ironically the next generation was to level the identical charge at Pearson: 'I have rather a dread', wrote R. A. Fisher to A. M. Walker in April 1940, 'of the procrustean process of forcing a problem to fit a favourite method; for, quite early in life, I was impressed by the fatal effects of this in Karl Pearson's work' (Bennett, 1990).

An American reviewer of *Natural Inheritance* noticed by Stigler, John Dewey, was more encouraging: 'It is to be hoped that statisticians working in other fields, such as the industrial and monetary, will acquaint themselves with Galton's development of new methods, and see how far they can be applied in their own fields'.

Stigler goes on to point out that when, three years later, Pearson published his book *The Grammar of Science* (see Pearson, 1937), 'he appeared to have forgotten Galton'. In his Chapter IV, 'Cause and effect – probability', Pearson treats the reader to an exegisis of Laplace's Bayesianism (as we should now say), in spite of listing both George Boole's *An Investigation of the Laws of Thought* and the third edition of John Venn's *The Logic of Chance* under 'Literature', both of which were very critical of the Bayesian position. That might be thought an irrelevant observation to our main theme were it not for the fact that Sir Harold Jeffreys, in a letter to Fisher dated 5 March 1938, considered his own Bayesian methods to be 'really based on *The Grammar of Science*' (Bennett, 1990), so that Pearson's book continues to exert a substantial indirect influence on modern statistics.

In practice Pearson's attitude to statistical inference, rather like Laplace's, was remarkably catholic, ranging from an acceptance of prior probabilities to the invention of his own famous chi-squared test of goodness-of-fit, which relies wholly on the concept of repeated-sampling (Pearson, 1900). This somewhat cavalier attitude to the niceties of inference led him into several difficulties, including controversy with Fisher about maximum likelihood and the misapplication of Bayes's model to the binomial prediction problem (Edwards, 1974, 1978). Galton, however, was entirely ignorant of the arguments about statistical inference, and since his most famous work assumed normality of the statistical distributions, the arguments would not have made much difference anyway.

E. S. Pearson, Karl's son and partial successor at University College, has recorded the changes his father made to the editions of *The Grammar of Science* in a useful introduction to the 'Everyman' edition published the year after Karl died in 1936 (see Pearson, 1937). The second, 1900, edition contained two new chapters on evolution, for help with which the author thanks Galton, W. F. R. Weldon and G. Udny Yule. Then in 1911 Part I of a third edition appeared with the subtitle 'Physical' and these two chapters were omitted as it was intended that they would be included in a new 'Biological' Part II (fated never to appear, partly because that was the year of Galton's death and his relatives had requested the undertaking of the massive biography). More importantly for us, this 1911 edition included a new chapter, 'Contingency and correlation – the insufficiency of causation', which may be examined for its references to Galton. Oddly enough there are none, not even in the 'Literature', which cites three papers by Pearson, a book by Elderton, and nothing else.

From these introductory remarks we see that the relationship between Galton and Pearson in matters of statistics was not as straightforward as might have been supposed, and that my title is not to be interpreted as signalling that Karl Pearson was the energetic conduit through which all Galton's statistical ideas reached the modern world. To understand their true relationship we must examine more closely what happened, and in its proper chronological order.

GALTON'S CONTRIBUTIONS TO MODERN STATISTICAL THEORY

I start by noting two preliminary matters. In 1872 Galton published 'Statistical inquiries into the efficacy of prayer', in which he gave possibly the first clear instructions for the statistical evaluation of a single factor, as follows:

> We must gather cases for statistical comparison, in which the same object is keenly pursued by two classes similar in their physical but opposite in their spiritual state; the one class being prayerful, the other materialistic. Prudent pious people must be compared with prudent materialistic people, and not with the imprudent nor the vicious (Galton, 1872).

I do not know whether this recommendation had the slightest impact, but it is certainly true that the statistical evaluation of factors one at a time held centre stage until Fisher's demonstration of the benefits of factorial analysis, in which several factors are handled simultaneously (see especially Fisher, 1935).

The second preliminary is the paper of 1874 with the Rev. H. W. Watson 'On the probability of the extinction of families', which is regarded as the origin of the theory of branching processes (Galton and Watson, 1874). There is, however, some doubt as to the directness of its influence. When Fisher examined the same problem in 1922 in connection with the extinction of genes he did not refer to the 1874 paper (Fisher, 1922), but one may certainly assume that he will read Galton's *Natural Inheritance* in which the matter is treated in Appendix F, and he may simply have regarded it (as he seems to have regarded most of Galton's contributions) as common intellectual property by the time he was writing. Besides, as Professor D. G. Kendall has pointed out in his histories of the matter (Kendall, 1966,

1975), there were continental and Russian antecedents as well. In addition the problem is one of probability rather than statistics, and therefore slightly outside our scope.

Galton's lasting contributions to the theory of statistics really start in the following year, 1875, with his gradual and highly original realisation that, as we should now say, a normal mixture of normal distributions is a normal distribution (Galton, 1875). Galton developed his famous 'quincunx' to provide an analogical proof of this statement, and the theorem itself was an essential preliminary to his insights into regression and correlation. Stigler, whose account cannot be bettered, says simply 'The crucial point is that Galton's conceptual use of the result was new and ingenious and represents the most important step in perhaps the single major breakthrough in statistics in the last half of the nineteenth century'.

Pearson, even in 1924 when commenting on this 1875 paper in Volume II of his *Life*, was not especially impressed. He placed his discussion in Chapter XIII, 'Statistical investigations, especially with regard to anthropometry', rather than with regression and correlation in Volume IIIa. After a rather critical discussion he concludes, almost apologetically, 'As we have endeavoured to show, the paper is extremely suggestive, but not every reader will be induced by its arguments to accept its conclusions'.

The next contribution to be noticed is the 1877 introduction of the idea of regression (or 'reversion' as Galton originally called it). Galton had engaged the services of seven friends to grow sweet peas from seeds which he had separated into seven groups by size. From the results he discovered that the variance in each group was sensibly the same, that the distributions were normal, and that the means of the groups were linearly related to the parental seed sizes *but with a slope of about one-third*: the offspring means were not the same as the parental means (which would have meant a slope of one) but were only one-third of the way from the overall mean towards the relevant parental mean. How was it, then, that in two successive generations the overall variance remained the same? The answer, in a nutshell, was that the sweet pea seeds were exhibiting the same phenomenon demonstrated by the quincunx: each group of parental seeds was producing a normal distribution of offspring, the variability of which in sum just compensated for the regression of the individual means to the overall mean (Galton, 1877). Exactly why the compensation was the right amount Galton would not discover for a little while; but he had understood the basic principle involved.

This work, in many ways the continuous analogue of Mendel's 1865 paper (which still lay buried in the libraries of Europe), is of timeless significance; what is more it could have been undertaken at any time since the dawn of science. But it fell to Francis Galton to plan the experiment and interpret the message.

For its reception we shall once again rely on Stigler's excellent account. Karl Pearson's first relevant paper did not appear until 1895 (Pearson, 1895; Morant, 1939), and it seems clear that he had not been following the Galtonian saga as it unrolled (he was, after all, a very young professor of applied mathematics and mechanics at University College, London at the time). But the older F. Y. Edgeworth certainly had been, for he and Galton were correspondents from 1881. Edgeworth's name is entirely missing from Pearson's *Life* save for its occurrence in a letter of Galton's to W. F. Sheppard in October 1892 referring to the need to find someone who would give a clear account of the use of the normal distribution in *Natural Inheritance*: 'Edgeworth has his own work and interests, and fails in sustained clearness of expression. He is moreover somewhat over fond of using higher and more mathematics than is always necessary. Watson is over busy and I think too fastidious and timid'. Prophetically Galton continued 'I have often considered what seems wanted and been very desirous of discovering someone who was disposed to throw himself into so useful and such high-class work. He might practically *found* a science, the material for which is now too chaotic'. In the following month Karl Pearson gave his first lectures on the laws of chance.

From 1881 onwards Edgeworth was very well read in statistics, and familiar with Galton's 1875 paper. In 1885 he read a paper, 'Methods of Statistics', at the jubilee meeting of the (Royal) Statistical Society in which he used the basic insight of Galton's work (Edgeworth, 1885). He went on to develop ideas from it which might have led to the analysis of variance had they been better understood at the time.

And it was in 1885 that Galton produced his third major advance, addressing the British Association on 'Regression towards mediocrity in hereditary stature'. Edgeworth was also a speaker at the meeting. In this paper (Galton, 1885, 1886), Galton presented his famous explanation of regression in terms of the geometry of the bivariate normal distribution, which he himself discovered by inspection of his two-way tabulation of data on the heights of adults and the mean heights of their parents (Figures 7.1 and 7.2). J. Hamilton Dickson of Cambridge provided the mathematical description necessary. Here at last was the explanation of the stability in variances generation by generation which was missing from the 1877 account.

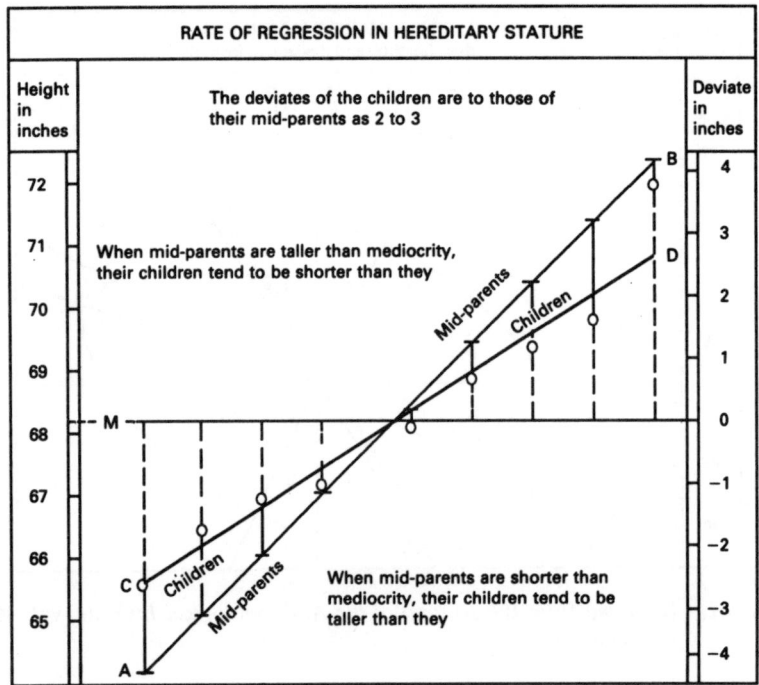

Source: Redrawn from the *Journal of the Anthropological Institute*, vol. 15 (1885) p. 248.

Figure 7.1 Regression in hereditary stature

All these advances Galton brought together early in 1889 with the publication of *Natural Inheritance* (Galton, 1889b). Pearson, as we have seen, was not impressed. Edgeworth, in a review in *Nature*, was. In May and June 1892 Edgeworth delivered the Newmarch Lectures at University College, and in the third and fourth lectures, 'The doctrine of averages' and 'Types and correlations', he discussed Galton's work. Pearson, a professor in the college, presumably attended (he certainly received a copy of the syllabus in advance, and in any case was already in correspondence with Edgeworth on questions of probability). Subsequently Edgeworth expanded his ideas on correlation in a series of papers, thus bringing recent developments to an even wider audience (Stigler, 1986).

When Karl Pearson did produce his first statistical publication, on asymmetric frequency curves in 1893 (Pearson, 1893), it was descended from Weldon and Edgeworth's interests, and was to initiate a long

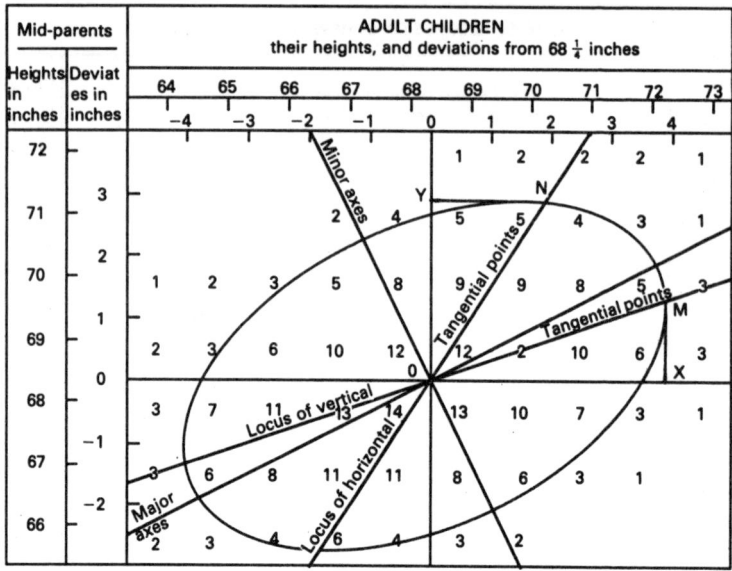

Source: Redrawn from the *Journal of the Anthropological Institute*, vol. 15 (1885) p. 249.

Figure 7.2 Galton's diagram of the geometry of regression, showing a typical elliptical probability-density contour and the two regression lines defined by the vertical and horizontal tangent points (Galton, 1886)

series of papers on non-normal distributions. The link with Galton's work was indirect.

The final important advance that Galton made was the notion of correlation itself, for he realised, after *Natural Inheritance* had gone to press, that if in his regression diagram the two variates were scaled so as to have the same probable error *on the paper* then both regression lines would have the same slope (each with respect to its proper axis), say *r*, which was therefore suitable as an 'index of co-relation', or 'correlation coefficient' as it soon became. Galton had discovered that regression to the mean and the association of two normal variables shared the same mathematical structure. These ideas were presented in December 1888 and published the following year (Galton, 1889a).

We may end our rapid survey of Galton's statistical work by noting his invention of normal probability paper in 1899, and its use to find the median (Galton, 1899).

This analysis of the years in which Galton was producing his most original statistical work shows that it was not through Pearson that the ideas were principally disseminated. Galton, who was sixty-seven when *Natural Inheritance* was published, was a well-known figure in the British scientific establishment, a Fellow of the Royal Society, a former president of the British Association, and a frequent contributor to *Nature*. To anyone of mathematical bent his ideas were simple enough. Given a few minutes' study of a bivariate normal distribution, regression and correlation are easy to understand. But it was Galton's genius which enabled him to understand them *first*. Insofar as the community of statisticians developed Galton's ideas, Edgeworth was the leader. Pearson busied himself with his skew curves; his great contribution to Galtonian studies was to be more administrative than academic. In the sphere of administration and organisation he attained lasting influence through his energy and leadership. The one idea of Galton's that he did develop was, of course, correlation, but correlation was to prove something of a diversion as far as modern statistics is concerned, a sort of circuitous and complicated detour on the road to the analysis of variance and covariance.

In this account the name of R. A. Fisher has hardly been mentioned so far because his well-known antipathy towards Pearson, who treated him roughly as a young man, might be thought to render unreliable any opinion he voiced about Pearson's role in the promulgation of Galton's methods. Yet he is such an important witness that his voice must be heard. Not only was he so much closer to the events than recent commentators, but his technical knowledge was more than adequate to ensure complete mastery of the earlier work. In addition he spent ten years of his life at University College as Galton Professor of Eugenics, and he was intimately involved with the first thirty years' work of the Eugenics Society.

In contrast to his comments on Pearson, Fisher's few references to Galton are straightforward and lacking in overtones. Fisher's historical interests, which were quite keenly developed in connection with figures like Gauss, Bayes, Mendel, Boole, Darwin, and Venn, did not extend to Galton, whom he seems to have regarded with amused affection, as indeed will anyone today who reads Galton's charming autobiography (Galton, 1908). He would probably have echoed Darlington's view that 'Galton as a man we can also see more clearly now than in his own time, more clearly indeed than from reading the volumes under which his devoted biographer has buried him. The fact is that the great discoveries of science often need a childlike character in the discoverer and Galton's simplicity was only too genuine' (see Galton, 1962).

Fortunately however, quite late in life, in 1951 well after the heat of the battle, Fisher did leave a little sketch in a few paragraphs headed 'Galton and Statistical Biology' which are worth quoting in full:

A man who, towards the end of the nineteenth century, played a peculiar part in precipitating modern developments was Francis Galton. A man of means and, had he chosen, of leisure, Galton made his name early in life as an African explorer. In 1869, evidently reacting eagerly to his cousin Charles Darwin's evolutionary theory, he had written *Hereditary Genius*, one of the most remarkable books of the century, and in it had demonstrated how apparently intangible concepts, at first vaguely apprehended, can be made quantitative and relatively precise by the collection and adequate presentation of statistical data. Throughout his life this possibility evidently exercised a fascination on his mind. In a crude way he attempted to collaborate in discussing the numerical results of his cousin's experiments with plants. He tried his hand at the statistical expression of meteorological phenomena, and, towards the end of his long life, armed with much experience, but without adequate mathematical technique, he became convinced that quantitative, and particularly statistical, methods were needed to consolidate Darwin's ideas, and to give confidence to their practical application. In Karl Pearson he found a man of boundless confidence and ambitious energy, and, with the sympathy of W. F. R. Weldon and his wide biological knowledge, Galton believed that a solid foundation could be built for a timely advance in the method and theory of biological research.

So far as Pearson's work is concerned, the immediate outcome was the appearance from 1894 onwards of a series of extensive memoirs entitled *Mathematical Contributions to the Theory of Evolution*. In all there were twenty-six of these, the first twelve appearing in the *Philosophical Transactions* of the Royal Society. The title chosen must be taken to represent rather Galton's hopes than Pearson's performance, for Pearson was here exploring his own general concepts in mathematical statistics, such as skew-frequency curves, contingency, and the rather numerous statistics to which, without distinction, he applied the term correlation. These developments were accompanied by extensive mathematical tables to facilitate their use. Mendel's laws are discussed in the twelfth memoir, but only to be dismissed as inadequate.

A more enduring consequence was the foundation, in 1901, of *Biometrika: A Journal for the Statistical Study of Biological Problems*. For many years this handsomely produced quarterly was undoubtedly the centre of development of mathematical statistics in this country. It accepted papers from outside Pearson's laboratory and included some of the most important advances of a period of rapid progress. In building up the high reputation of this journal Pearson's labours as editor were constant and indefatigable and constituted the greater part of the scientific activity of his later life.

Although in following his own bent Pearson undoubtedly wandered far from Galton's intention, yet he may be regarded as ploughing the ground in preparation for later developments. That the huge mass of his writings have now little value must be ascribed to two circumstances – first, that his mathematics were on the whole clumsy and lacking in penetration and, second, that without the power of self-criticism he was unable and unwilling to correct his numerous errors or to appreciate the work of others, which would have been of the greatest assistance. He seems to have regarded observational material principally as a means of illustrating his *a priori* concepts, not as a means of correcting them or as providing problems of interpretation in which statistical methods might be of service. He seems to have felt a contempt for the work previously done in the theory of errors and to have known little about it. He certainly regarded the skew-frequency curves he invented as improvements to be substituted for the normal law of errors (Fisher, 1951).

I have the impression that Fisher took great care in writing these paragraphs, and has given credit where credit is due. The only mistake, it seems to me, is in suggesting that Pearson knew little about previous work in the theory of errors, and the only serious omission is any mention of the justly famous goodness-of-fit paper of 1900. That must have been simple oversight, however, because in the later editions of *Statistical Methods for Research Workers* Fisher included a historical note in which he said of Pearson's chi-squared test 'This, I believe, is the great contribution to statistical methods by which the unsurpassed energy of Prof. Pearson's work will be remembered' (Fisher, 1946).

THE GALTON PROFESSORSHIP AND LABORATORY

Karl Pearson wielded enormous influence through his activities at University College, so many of them inspired and financed by Galton himself. The Biometric Laboratory, the Eugenics Laboratory, the Galton Professorship of National Eugenics (later Human Genetics), the Professorship of Statistics, the *Annals of Eugenics* (later *Annals of Human Genetics*), *Biometrika*, . . . the list is endless. This was Pearson's true contribution to the furtherance of Galton's ideals, but, as we have seen, not to the furtherance of Galton's ideas. Indeed the absolute dominance of Pearson in the British statistical world in the early years of this century seems now to me to have actually slowed the development of Galton's statistical ideas. Pearson's emphasis on skew curves and correlation coefficients surely hindered the development of the analysis of variance, and his antagonism towards Mendelism certainly held back the theoretical explanation of the correlations between relatives. (See the account of the Galton Laboratory in Chapter 14).

GALTON'S INTELLECTUAL HEIR?

In conclusion, I venture to advance a thesis: simply that in statistics the young R. A. Fisher was the real inheritor of Galton's intellectual mantle. Perhaps this explains something of the antagonism Pearson felt towards him. This thesis stands principally on three of Fisher's early contributions. First, his 1918 paper 'The correlation between relatives on the supposition of Mendelian inheritance', which brilliantly synthesised the biometrical and Mendelian standpoints (Fisher, 1918a); secondly, his contemporaneous invention of the analysis of variance (the phrase first occurs in his *Eugenics Review* paper 'The causes of human variability' [Fisher, 1918b]), which was not only Galtonian in its context but manifestly starts where Galton's dissection of normal variation had left off; and thirdly, his pathbreaking determination of the exact sampling distribution of the Galtonian correlation coefficient, which heralded the dawn of modern mathematical statistics (Fisher, 1915).

   In comparison with these advances, Pearson's own intellectual contribution now seems tangential. How little he was able to appreciate the direction in which the future lay may be illustrated by the well-known fact that he was one of the Royal Society's referees whose reports led to the rejection of Fisher's 1918 paper (the other was R. C. Punnett, the first Arthur Balfour Professor of Genetics at Cambridge; in telling the story Fisher used to say 'I later succeeded both of them'. See also Norton and Pearson, 1976). In another instance (Bennett, 1983) Pearson refused to publish in *Biometrika* Fisher's paper 'On the "probable error" of a coefficient of correlation deduced from a small sample' (Fisher, 1921), in which Fisher got out the exact sampling distribution for the intraclass coefficient (complementing his earlier work on the interclass coefficient).

We do not have to look far for the personal link between Galton and Fisher. Leonard Darwin, Charles Darwin's fourth son, was well acquainted with both men, the one twenty-eight years his senior and the other forty years his junior. He succeeded Galton as president of the Eugenics Education Society in 1909 and served for eighteen years, lending advice and encouragement to the young Fisher throughout. *The Genetical Theory of Natural Selection* (Fisher, 1930) is dedicated to him, and it is he whom Fisher thanks for suggesting the topic of the 1918 paper: 'Finally, it is a pleasure to acknowledge my indebtedness to Major Leonard Darwin, at whose suggestion this inquiry was first undertaken, and to whose kindness and advice it owes its completion' (Fisher, 1918a).

As J. H. Bennett relates in his fascinating introduction to the Fisher–Darwin correspondence (Bennett, 1990), Darwin's earliest letters, from 1915, set out various problems arising from Galton's work which he hoped Fisher would solve. Leonard Darwin had already encountered difficulty with Pearson; in 1913 he had questioned the use of the phrase 'the relative influence of heredity and environment' (Darwin, 1913 – reprinted in full as an appendix to this chapter), to which Pearson had replied by means of a paper in *Biometrika* with the astonishing title 'On certain errors with regard to multiple correlation occasionally made by those who have not adequately studied the subject' (Pearson, 1914). When he came to examine the controversy two years later, Fisher thought Darwin had been right, but Darwin, who had inherited a horror of controversy from his father, advised him not to argue with Pearson (Bennett, 1983). From then on Fisher's own analysis developed rapidly, culminating in the 1918 paper.

Bennett (1983) remarks that 'For Fisher the friendship with Darwin, with his close links with Charles Darwin and Francis Galton, had special significance'. After Darwin's death in 1943 Fisher wrote of his 'very dear friend' that he was 'surely the kindest and wisest man I ever knew' (Bennett, 1983).

Appendix

The following extract is reprinted from *Eugenics Review*, vol. 5 (1913) pp. 153–4.

CORRESPONDENCE.

HEREDITY AND ENVIRONMENT.

I never read any general work on Eugenics without being tempted to make one broad criticism, a criticism which, however, I make with some fear and trembling, because of its wide application. In all of them such expressions as 'the relative influence of heredity and environment' are of common occurrence; but whenever read by me such phrases create a sensation of discomfort from the conviction that I am unable to attach a definite meaning to them. It is impossible to compare heredity as a whole with environment as a whole as far as their effects are concerned; for no living being can exist for a moment without either of them. Moreover, in order to compare two things so as to be able to use the words more or less in connection with such a comparison, we must have a common unit of measurement applicable to them both. But what is the unit by which both heredity and environment may be measured? I, myself, have no idea. May we not be discussing questions as illogical as enquiring what portion of the area of a rectangle is due to its width and what to its length? Is it ever wise to use words in scientific literature without endeavouring to attach a definite meaning to them?

Perhaps those learned in the statistical examination of eugenic questions may reply somewhat as follows, though, doubtless, in neater phraseology. Take any quality, and we find that the human beings composing any community differ more or less considerably as regards that quality. Now we can measure the correlation between the differences shown in this quality and the differences of environment to which the members of the community in question had previously been exposed. This is one correlation. Then we can also measure the correlation co-efficient between, say, father and son, as regards the quality in question. Here is a second correlation; and if we are told that the relative influence of environment and heredity is measured by the ratio between these two correlation co-efficients, we certainly do thus get a clear conception of what is meant. But putting aside the fact that this is not a meaning which springs very readily from the words in question, we must ask whether the ratio thus indicated is one which is likely in all circumstances to serve as a safe guide to the practical eugenist. Surely what we want to know is how we can do most good – whether by attending to reforms intended to affect human surroundings, or to reforms intended to influence mankind through the agency of heredity. But does this ratio give us any sure indication of the relative amount of attention which should be paid to these two methods of procedure? This is a question which, it seems to me, has not yet received sufficient attention in connection with the statistical study of eugenics.

Imagine an ideal republic, in some respects similar to that designed by Plato, where not only were all the children removed from their parents, but where they were all treated exactly alike. In these circumstances none of the differences between the adults could have anything to do with the differences of environments, and all must be due to some differences in inherent factors. In fact, the environmental correlation co-efficient would be nil, whilst the hereditary correlation co-efficient might be high. But the environment in this imaginary community might be wretched and easily improvable, whilst circumstances might make eugenic reform temporarily impossible. Would not he be a very foolish statesman who, in any circumstances at all like these, would be deterred from devoting his first efforts to environmental reform by being told that the environmental co-efficient was very small? Or, again, is it not at least conceivable as regards some zymotic disease, that the differences of environment to which individuals are exposed in no way affect the liability to catch the disease in question, all being equally exposed to infection; whilst, at the same time, might it not be possible that by some general change of environment the microbe could be exterminated and the disease stamped out? Here, again, everything needed could be accomplished by an environmental reform, in spite of the environmental co-efficient of correlation being nil. Do not such possibilities, fanciful though they be, make one wish for more light on the subject?

My suggestion is that at present we should, as far as possible, avoid such phrases as the relative influence of heredity and environment, whilst always holding in view the relative possibilities of doing good by attending to heredity and to environment. To state the problems to be solved in this form would not, I believe, make the statistical method in the least less essential; and the result would be all the more useful, from thus more clearly keeping in view the goal of our efforts. Moreover, the importance of the heredity factor would in the end thus be made to stand out more clearly because more intelligibly. On the other hand, if, as is quite possible, I am wrong, and the statistician is able to produce a completely destructive reply to these criticisms, then they will have fully answered their purpose in bringing forth that reply.

LEONARD DARWIN.

References

Bennett, J. H. (ed.) (1983) *Natural Selection, Heredity, and Eugenics. Including selected correspondence of R. A. Fisher with Leonard Darwin and others* (Oxford: Clarendon Press).

Bennett, J. H. (ed.) (1990) *Statistical Inference and Analysis: Selected Correspondence of R. A. Fisher* (Oxford: Clarendon Press).

Darwin, L. (1913) 'Heredity and environment', *Eugenics Review*, vol. 5, pp. 153–4.

Edgeworth, F. Y. (1885) 'Methods of statistics', *Jubilee Volume of the Statistical Society*, pp. 181–217.

Edwards, A. W. F. (1974) 'The history of likelihood', *International Statistical Review*, vol. 42, pp. 9–15.

Edwards, A. W. F. (1978) 'Commentary on the arguments of Thomas Bayes', *Scandinavian Journal of Statistics*, vol. 5, pp. 116–18.

Fisher, R. A. (1915) 'Frequency distribution of the values of the correlation coefficient in samples from an indefinitely large population', *Biometrika*, vol. 10, pp. 507–21.

Fisher, R. A. (1918a) 'The correlation between relatives on the supposition of Mendelian inheritance', *Transactions of the Royal Society of Edinburgh*, vol. 52, pp. 399–433.

Fisher, R. A. (1918b) 'The causes of human variability', *Eugenics Review*, vol. 10, pp. 213–20.

Fisher, R. A. (1921) 'On the "probable error" of a coefficient of correlation deduced from a small sample', *Metron*, vol. 1, pp. 3–32.

Fisher, R. A. (1922) 'On the dominance ratio', *Proceedings of the Royal Society of Edinburgh*, vol. 42, pp. 321–41.

Fisher, R. A. (1930) *The Genetical Theory of Natural Selection* (Oxford: Clarendon Press).

Fisher, R. A. (1935) *The Design of Experiments* (Edinburgh: Oliver and Boyd).

Fisher, R. A. (1946) *Statistical Methods for Research Workers*, 10th edn (Edinburgh: Oliver and Boyd).

Fisher, R. A. (1951) 'Statistics', in *Scientific Thought in the Twentieth Century*, A. E. Heath (ed.), pp. 33–55. London: Watts.

Fisher, R. A. (1971–4) *Collected Papers of R. A. Fisher*, J. H. Bennett (ed.), vols I to V. (University of Adelaide).

Forrest, D. W. (1974) *Francis Galton: The Life and Work of a Victorian Genius* (London: Elek).

Galton, F. (1872) 'Statistical inquiries into the efficacy of prayer', *Fortnightly Review*, vol. 12, pp. 125–35 (reprinted by the Eugenics Society in *Sir Francis Galton: Three Memoirs*, 1951).

Galton, F. (1875) 'Statistics by intercomparison, with remarks on the law of frequency of error', *London, Edinburgh and Dublin Philosophical Magazine*, 4th series, vol. 49, pp. 33–46.

Galton, F. (1877) 'Typical laws of heredity', *Nature*, vol. 15, pp. 492–5, 512–14, 532–3.

Galton, F. (1885)'Regression towards mediocrity in hereditary stature', *Journal of the Anthropological Institute of Great Britain and Ireland*, vol. 15, pp. 246–63.

Galton, F. (1886) 'President's Address', *Journal of the Anthropological Institute of Great Britain and Ireland*, vol. 15, pp. 489–99.

Galton, F. (1889a) 'Co-relations and their measurement, chiefly from anthropological data', *Proceedings of the Royal Society of London*, vol. 45, pp. 135–45.

Galton, F. (1889b) *Natural Inheritance* (London: Macmillan).

Galton, F. (1899) 'A geometric derivation of the median value of a system of normal variants, from two of its centiles', *Nature*, vol. 61, pp. 102–4.

Galton, F. (1907) *Probability, the Foundation of Eugenics* (Herbert Spencer Lecture) (Oxford: Clarendon Press).

Galton, F. (1908) *Memories of My Life* (London: Methuen).

Galton, F. (1962) *Hereditary Genius* Fontana Library edition, with an Introduction by C. D. Darlington (London: Collins).

Galton, F. and H. W. Watson (1874) 'On the probability of the extinction of families', *Journal of the Anthropological Institute of Great Britain and Ireland*, vol. 4, pp. 138–44.

Hacking, I. (1990) *The Taming of Chance* (Cambridge University Press).

Kendall, D. G. (1966) 'Branching processes since 1873', *Journal of the London Mathematical Society*, vol. 41, pp. 385–406 (reprinted in Kendall and Plackett, 1977).

Kendall, D. G. (1975) 'The genealogy of genealogy: branching processes before (and after) 1873', *Bulletin of the London Mathematical Society*, vol. 7, pp. 225–53.

Kendall, M. G. and R. L. Plackett (eds) (1977) *Studies in the History of Statistics and Probability*, vol. II (London: Griffin).

MacKenzie, D. A. (1981) *Statistics in Britain 1865–1930: The Social Construction of Scientific Knowledge* (Edinburgh University Press).

Morant, G. M., with the assistance of B. L. Welch (1939) *A Bibliography of the Statistical and Other Writings of Karl Pearson* (London: Biometrika Office, University College).

Norton, B. and E. S. Pearson (1976) 'A note on the background to and refereeing of R. A. Fisher's 1918 paper "The correlation between relatives on the supposition of Mendelian inheritance"', *Notes & Records of the Royal Society of London*, vol. 31, pp. 151–62.

Pearson, E. S. (1990) *'Student': A Statistical Biography of William Sealy Gosset*, edited and augmented by R. L. Plackett with the assistance of G. A. Barnard (Oxford: Clarendon Press).

Pearson, K. (1893) 'Asymmetrical frequency curves', *Nature*, vol. 48, pp. 615–16.

Pearson, K. (1895) 'Note on regression and inheritance in the case of two parents', *Proceedings of the Royal Society of London*, vol. 58, pp. 240–1.

Pearson, K. (1900) 'On the criterion that a given system of deviations from the probable in the case of a correlated system of variables is such that it can be reasonably supposed to have arisen from random sampling', *London, Edinburgh and Dublin Philosophical Magazine*, 5th series, vol. 50, pp. 157–75.

Pearson, K. (1914) 'On certain errors with regard to multiple correlation occasionally made by those who have not adequately studied the subject', *Biometrika*, vol. 10, pp. 181–7.

Pearson, K. (1914–30) *The Life, Letters and Labours of Francis Galton*, vol. I (1914): 'Birth 1822 to Marriage 1853'; vol. II (1924): 'Researches of Middle Life'; vol. III^A (1930): 'Correlation, Personal Identification and Eugenics'; vol. III^B (1930): 'Characterisation, Especially by Letters'; Index (Cambridge University Press).

Pearson, K. (1937) *The Grammar of Science*, Everyman's Library edition, with an Introduction by E. S. Pearson (London: Dent).

Porter, T. M. (1986) *The Rise of Statistical Thinking, 1820–1900* (Princeton University Press).

Stigler, S. M. (1986) *The History of Statistics: The Measurement of Uncertainty before 1900* (Cambridge, Mass: Belknap Press).

8 Galton on Human Growth and Form

J. M. Tanner

Francis Galton's first study of growth and form (or anyway the latter) was made with a sextant and trigonometric tables on a beautiful but apparently unapproachable Hottentot lady who was standing under a tree at a mission station in South West Africa in 1850. She was the wife of one of the Hottentot interpreters. Galton records the incident thus:

> I profess to be a scientific man, and was exceedingly anxious to obtain accurate measurements of her shape; but there was a difficulty in doing this. I did not know a word of Hottentot, and could never therefore have explained to the lady what the object of my footrule could be; and I really dared not ask my worthy missionary host to interpret for me. The object of my admiration stood under a tree, and was turning herself about to all points of the compass, as ladies who wish to be admired usually do. Of a sudden my eye fell upon my sextant; the bright thought struck me, and I took a series of observations upon her figure in every direction, up and down, crossways, diagonally, and so forth, and I registered them carefully upon an outline drawing for fear of any mistake: this being done, I boldly pulled out my measuring tape, and measured the distance from where I was to the place where she stood, and having thus obtained both base and angles, I worked out the results by trigonometry and logarithms (Galton, 1853).

After this promising beginning, however, Galton's interest in the human body faded and was revived only in the context of his overriding life-interest, 'viviculture'. In 1873 Galton published in *Fraser's Magazine* an article entitled 'Hereditary Improvement' which he himself called 'audacious' in a letter to the Swiss scientist Alphonse de Candolle. In it he proposed the desirability of creating a superior caste of gifted individuals, drawn from all social classes but interbred so that eventually they would become the most influential group in the nation (surely the model of Plato's Guardians could not have been far from Galton's mind). Indeed so influential would they become that eventually legislation would be passed (this is in England in 1873 remember) preventing heirs deficient in natural gifts from inheriting vast fortunes. These guardians would live under better conditions than others, multiply more rapidly and 'treat their compatriots with all kindness, so long as they [the compatriots that is] maintained celibacy'.

First, however, find your guardians. For this it is necessary to initiate a stocktaking of the nation as a whole. Already in *Hereditary Talent and Character* (1865) – where these ideas of viviculture are already clearly present – Galton calls it 'a great and common mistake . . . to suppose that high intellectual powers are generally associated with puny frames and small physical strength . . . most great men are vigorous animals with exuberant powers . . . there is no reason to suppose that in breeding for the highest order of intelligence we should produce . . . a feeble race' (1865, p. 164). So in 1873 Galton proposed a national stocktaking which included 'all about their respective health and strength and constitutional vigour; to learn the amount of a day's work of men in different occupations; their intellectual capacity, so far as it can be tested at schools; . . . [and] sanitary questions . . . to give a correct idea of the present worth of our race, and means of comparison some years hence of our general progress or retrogression' (1873, p. 125).

GALTON ON GROWTH

Accordingly in the following year (1874) Galton approached the recently formed Anthropological Institute (AI) with a 'Proposal to Apply for Anthropological Statistics from Schools' (Galton, (1874a)). Considering that schoolmasters were 'trustworthy and intelligent in no small degree' the statistics would be reliable, and the boys measured would grow up with favourable recollections of the procedure and hence would willingly submit to further measurement when grown up and working in universities, factories, and so on (Galton clearly had in mind his family record study, initiated ten years later). Galton also observed that such a study would 'give us the law of growth in different classes, both in town and country. This is known to vary exceedingly under different conditions, but exact numerical determinations have yet to be established' (1874a, p. 310). In the discussion which followed, Galton, it is reported, made clear his desire to obtain statistics from 'schools of all description, such as Public Schools, middle class schools and others, down to those of pauper children' (1874a, p. 311).

Also in the discussion Serjeant Cox said that Quetelet, 'in his recent work [that is *Anthropométrie*] had made an extensive and valuable collection of statistics . . . of this kind which would form a basis for comparison'. To this remark Galton did not reply, presumably because he was not acquainted with the reference. He admired Quetelet, that most brilliant star of the Belgian sky (see Tanner, 1981, pp. 122–41) and thought his *Letters . . . on the Theory of Probabilities*, translated into English in 1849, the best introduction to statistics for anthropologists: he said so in his article 'Statistics' in the British Associations's handbook entitled *Notes and Queries on Anthropology* (1872). Indeed this small book of Quetelet's, which grew out of the lessons he gave in the period around 1837 as private tutor to the two princes of Saxe-Coburg, Ernest and Albert, later Prince Consort, is cited by Karl Pearson (1914–30) as Galton's first – and immensely important – introduction to the Laplace–Gauss curve of variation (that's what Karl Pearson calls it, though it is actually a curve of error, of course) (Pearson, 1924, vol. 2, p. 89).

There is no evidence in Galton's writings, however, that he was familiar with Quetelet's authoritative *Sur l'homme et le développement de ses facultés* (1835, translated 1842; in later edition renamed *Physique Sociale*) though he had evidently (see Galton, 1875) read, at least later, his *Anthropométrie* (1870). *Physique Sociale*, despite being one of the most influential books of the century in relation to the founding of statistics, was described by Florence Nightingale, a great admirer of Quetelet's, as unobtainable in England in 1872 (see Tanner, 1981, p. 141).

More strangely Galton seems to be unaware at this time of the work – of the existence even – of Charles Roberts and the factory commissioners, who had just completed a massive survey of working-class children's growth (see below).

Galton's proposal was in the event only taken up by a number of public schools, five in the country (Marlborough, Clifton, Hailey-bury, Wellington and Eton Colleges) and four in towns (City of London School, Christ's Hospital, King Edward's School Birmingham and Liverpool College). Marlborough College had both a full-time biology master and a resident doctor, and these two made their returns in the form of a paper to the Anthropological Institute (Fergus and Rodwell *in* Galton, 1874b). They measured height and weight, and chest, upper arm and head circumferences on 550 boys aged ten to nineteen. They – or perhaps Galton himself, it would have been characteristic of him – constructed a height measuring instrument far superior to anything in use at that time. It had a counter-weighted headboard sliding between vertical guides, a design not reinvented (by R. H. Whitehouse and myself) until the 1950s and since then the standard.

Galton (1874b) wrote a brief comment on this paper and in the following year (Galton, 1875) analysed the heights of fourteen-year-olds using a graphical method of estimating the 8th, 25th, 50th and 75th and 92nd percentiles (that is ± 1 and ± 2 probable errors). He compares his method with that used by Quetelet in *Anthropométrie* (1870). This is the paper that introduced the system of percentiles for characterising a distribution, which is one of the two lasting contributions of Galton to the field of auxological analysis. (At this juncture the word 'percentile' was not used; that was introduced, it seems, nearly ten years later in a table of the results of the Anthropometric Laboratory [see below] in 1884 [Galton, 1884b: see also Galton, 1885]).

Galton then analysed the difference in height between boys in the five country schools and those in the town schools. Not surprisingly he found the country-school boys to be taller than those in the towns by some 3.5 cm. He ignores however the rather obvious fact that the parents of the country-school boys were of a higher social class than those of the town-school boys. He does make the interesting comment – based on poor statistics but seeming to show he knew more about growth than he mostly lets on – that at fifteen years the two means are closer together, so some of the difference is due to 'retardation in growth', some to 'total suppression'. Galton had always believed strongly in the evil influences of town life and at this juncture he writes something nearly Lamarkian: that the smaller size is due to the parents having grown up in the towns, and perhaps the grandparents also.

FRANCIS GALTON AND CHARLES ROBERTS

There is an intriguing byway of history at this time in the relations between Galton and a more substantial figure in the area of growth – Charles Roberts (see Tanner, 1981, pp. 172–80). Roberts was born in Yorkshire, missed out on Oxbridge, had a medical education of some brilliance at St George's Hospital in London and wound up as one of the five doctors involved in examining and measuring some 10 000 children, of both sexes, working in the textile factories or similar establishments in Yorkshire and Lancashire. This immense survey, which included heights and weights, was carried out in 1872–3.

Yet Galton gives no sign of ever having heard of it, and this is all the more strange because this survey was a successor to the Factory Commission Surveys of 1833 and 1837, surveys in which the dominant figure was Leonard Horner (Tanner, 1981, pp. 153–61). Leonard Horner (1785–1864) was a close friend of Francis Galton's father, Tertius, and in his youth Francis was a frequent visitor to the Horner's house in Gower Street. Horner was a passionate factory commissioner until 1855, and it is inconceivable that Francis Galton had not heard him talk about the use of children's measurements both to check their alleged ages (for entrance to factories) and to monitor their health and conditions of work. The results of the 1872–3 survey were published in Parliamentary Papers in 1873 and Roberts made an exemplary analysis of the data in the following year in *St George's Hospital Reports* (Roberts, 1874–6). Roberts had made an in-depth study of Quetelet's work – he was its main expositor and critic in the 1870s – and laid as great a stress as Galton on the variation between different children of the same age.

Which way the influence went (if indeed either way) is unclear. Galton, and presumably the worthies of the Anthropological Institute, remained in ignorance of the factory commissioners' work; it was Roberts who contacted Galton when he heard, through his medical colleague Fergus at Marlborough College, what was going on. He then got permission to include the Anthropological Institute records in his 1874–6 paper and his later *Manual of Anthropometry* (1878). Roberts was in constant communication with Henry Bowditch (1840–1911) (Tanner, 1981, pp. 185–96) in Boston and it was he who introduced Bowditch to frequency distributions and Galton's system of percentiles. Subsequently, in 1891, Bowditch produced the first centile standard charts for children's heights and weights at successive ages (Tanner, 1981, p. 195). Neither Roberts nor Galton left successors in the field of growth in the UK; but Bowditch was the instigator of a considerable school in the USA, which included W. T. Porter and, a little later, Franz Boas, generally regarded as the father of work in modern auxology.

Thus Galton's chief – indeed only – legacy in the field of growth, that is the method of percentiles, was transmitted through Roberts. Galton and Roberts actually worked together in the early 1880s on the Anthropometric Committee of the British Association (BAAS). The Committee ran from 1875 to 1883 and issued annual reports. Galton was throughout a member; Roberts only joined in 1879 but thereafter played an increasingly prominent part and wrote much of the extensive final report. For nearly ten years Galton – member of the rich Victorian upper class, a man who left just £200 to Gifi, his life-long devoted valet – and Roberts, the quiet scholarly middle-class Yorkshire doctor, must have worked, if not hand in hand, then back to back. Yet Charles Roberts is nowhere mentioned in Karl Pearson's massive biography.

Though Galton did not deal subsequently with any more growth statistics, he never lost his interest in the potential of growth data for monitoring health. At the London Congress of the Royal Institute of Public Health he gave an address on Anthropometry in Schools (Galton, 1906): 'Anthropometry furnishes the readiest method of ascertaining whether a boy [*always* a boy, never a girl!] is developing normally or otherwise, and how far the average conditions of pupils at one institution differs from those at others'. He continued, presciently, 'No programme of anthropometry in any school can be considered complete unless it provides for the collection of data during the after-lives of the pupils.' He suggested, with typical White-Knight (whom he rather resembled) common-sense that the follows-up should take place each 29 February and this day school gatherings should celebrate 'the works of living men who were formerly pupils, but then engaged in the battle of life. Their doings would be spoken of and hearty sympathy evoked'.

THE ANTHROPOMETRIC LABORATORY

Galton tried to establish anthropometric laboratories in the above and other schools, where not only physical measurements but also – and predominantly – physiological and psychometric measurements should be taken. However in this he failed. In a later version of this proposal (*Fortnightly Review*, 1882) he added a medico–metric section where GPs could send patients for check-ups. In line with this, when an international exhibition was held in South Kensington in 1884 he set up, at his own expense, an anthropometric laboratory.

Visitors, who paid 3d for the privilege, passed along an area 36 feet by 3 feet, being tested as they progressed. Height, sitting height and weight were taken, arm span, vital capacity, strength of pull and grip, swiftness of a blow, reaction time, acuity of vision and hearing, discrimination of colour and judgement of length (Galton, 1884a). There was difficulty with shoes, so height was measured with them on and the height of the shoe heel subtracted (not really a valid procedure as posture is altered by the shoe). Nearly 10 000 persons were measured, the great majority adults. When the exhibition ended in 1885 the laboratory was moved to the nearby South Kensington Museum where it continued to operate, with a permanent technician, for the space of eight years. In this laboratory head measurements were also taken, together with fingerprints.

Most of the visitors were, of course, middle and upper class. Galton (1884b) gave his results for height, sitting height, span, strength of grip, and so on in percentiles, citing the 95th, 90th, 80th, 70th, 60th and 50th, and those symmetrically below. The numbers of subjects were 881 and 770 for men and women for stature, 520 and 276 for weight, 579 and 276 for grip, but Galton writes that he 'did not care to have the records exhausted' so took 'as many as seemed in each case to be sufficient to give a trustworthy result'. Galton discusses the overlap of the distributions for men and women, noting that the overlap for strength is considerably less than that for height (a theme taken up by a contemporary writer in *Punch* who made fun of the strongest woman at the exhibition, whose grip amounted to 86 lb which was about average for a man). This 1884 paper incidentally contains the only remarks Galton made about the form, or shape, of the body. He considers the ratio of sitting height to stature (a very usual item in contemporary physical anthropology) and notes that in men 'a moderate increase in tallness is not associated with a disproportionate increase of length of legs' (that is a decrease in ratio) whereas in women it is. This was not a subject that Galton thereafter pursued.

Karl Pearson claimed that the Anthropometric Laboratory 'gave a vigorous push to Anthropology'. In 1894, when the laboratory closed, the equipment went to Oxford University, and in Cambridge and Dublin copies of the laboratory were set up. So 'pushed' it may have been, but the history of physical anthropology in the UK shows it was still insufficient to bring activity in that field to the level it achieved in many European countries and in the USA.

THE RECORD OF FAMILY FACULTIES AND THE INHERITANCE OF STATURE

Galton's second (after percentiles) and most lasting contribution to the study of growth and form was his demonstration that the inheritance of stature could be studied by mathematical means. In January 1884, when the Anthropometric Committee of the BAAS had just been wound up after a run of eight years, and the Anthropometric Laboratory was in the offing, Galton suggested, in letters to the *Times* and *Nature*, that records of family likeness could be obtained from a questionnaire distributed to members of the public, with prizes awarded to those who provided the most adequate answers. Very large numbers of questions were asked (Galton, 1884c) amongst which were those relating to height, colour of eyes and hair, and bodily strength.

Galton only obtained 110 responses to his questionnaire, and he distributed the money amongst 85 of the responders. However these questionnaires contained information on the reported heights of 930 adult offspring and their 205 parents. The main paper reporting these results was issued in 1885 and the results were largely repeated in Galton 1886, together with results on the variation in stature amongst brothers obtained by a separate, more personal inquiry.

Galton first transformed all his female heights into male equivalents by multiplying by 1.08 (which is entirely in line with what we should do nowadays). He called the mean of father's and transformed mother's height 'mid-parent height' (present practice tends to take the mean of father and untransformed mother, so on average Galton's mid-parent heights are 6.5 cm greater than modern ones (Tanner, 1986). Galton then tabled the distributions of differences of offspring's heights from mid-parents' heights according to five classes of mother–father difference (thus very tall fathers and short mothers, and averagely tall both). He found no indications of an effect of such difference. Galton examined the data roughly for assortative mating and found no evidence of it, and said he had insufficient data to permit an examination of whether there was a closer height relationship between parent and offspring of the same or different sex.

As is well known, Galton found that the height deviate from the median of the offspring was two-thirds that of the deviate of the mid-parent from the mid-parent median (see Chapter 7 of this book). This is the paper in which the word 'regression' (to the average) is used (the regression of 0.67 is lower than that found in modern data, presumably because of errors in the reported measurements. Galton's later paper gives the correlation coefficient as 0.47, which contrasts with a typical value of about 0.63 nowadays [Tanner, 1986]).

CO-RELATION

Galton's anthropometric data also led to that marriage of ideas of variation and regression that produced the correlation coefficient, discussed elsewhere in this volume. Galton puzzled over how to relate stature, sitting height, arm length and so on, when each was measured on a different scale. In 1888 he suddenly realised that if each was expressed in terms of its own variability, then the regression coefficient would be a measure of the relationship between them. In his introduction to this paper (Galton, 1888a) he uses the relation between the length of the arm and the length of the leg as his example. Thus the model of anthropometry underlay what is probably the greatest single contribution Galton made to science.

COMPOSITE PICTURES

We now turn briefly to Galton's pioneering work in combining portraits of a number of people into a composite picture, in which individual peculiarities disappeared and the general form of the human face – the face of Quetelet's beloved *homme moyen* – emerged. He did this by simply exposing on one photographic plate, say, ten portrait photographs, all of the same size, each for one-tenth the time of a normal exposure. After some experimentation very satisfactory results were obtained; results, that is, which resembled a perfectly real person, asymptotically the *homme moyen* himself (Galton, 1878). As an example he combined portraits (mug-shots) of numbers of criminals convicted of three different sorts of crimes: murder, forgery and sexual offences. Perhaps to his disappointment the generalised murderer and the generalised forger looked remarkably like the generalised prison warder: 'The common humanity that underlies them has prevailed' he said. He later tried to distinguish in the same way patients with tuberculosis from those with a variety of other diseases, again with little usable result (Galton and Mohamed, 1881).

Galton made no effort to measure these portraits and the idea that this was an early form of photogrammetry is mistaken.

GALTON'S LEGACY ON GROWTH AND FORM

What then is Galton's legacy in the field of human growth and form? In growth, one thing only: percentiles, that representation of an individual's status with regard to his fellows. Derived from Quetelet, transmitted by Roberts, it was first used in growth tables in 1891 by Bowditch. It has been used ever since; nearly all standards of height and weight of children are couched in terms of percentiles (or as all auxologists now call them 'centiles'). Even though standard deviation scores would do as well for height measurements, centiles are used because they can be immediately understood by parents. The statement that 'twenty per cent of all healthy boys are shorter than your son; eighty per cent are taller' – a statement in essentials first made by Galton – conveys at once the answer to the question 'is my son abnormal?'

It is true that the Galton-inspired data on middle-class children's heights were a powerful addition to the data on working children collected by Roberts. The combination – published by Roberts, not Galton – was one of the first substantial studies of the effect of social class on growth. Galton himself was not much interested, it seems, in epidemiological auxology, only in the auxology of the élite.

In the study of human shape or form, Galton's contribution was again a statistical tool, a tool of widespread applicability to science as a whole, but derived strictly from considering how to convey the notion that the length of men's arms and legs tended to vary in concert, whereas their heights and head circumferences went their separate ways. Thus the correlation coefficient was born, and underlying it, derived also from anthropometric considerations, was the regression coefficient. These two are the glories of the Galton legacy.

References

Galton, F. (1853) *Tropical South Africa* (London: John Murray), second edition, under the title *Narrative of an Explorer in Tropical South Africa* (1889) (London: Ward, Lock and Co).

Galton, F. (1865) 'Hereditary talent and character', *Macmillan's Magazine*, vol. 12, pp. 157–66.

Galton, F. (1872) 'Statistics', in: *Notes and Queries on Anthropology* (British Association for the Advancement of Science).

Galton, F. (1873) 'Hereditary Improvement', *Fraser's Magazine*, vol. 7, pp. 116–30.

Galton, F. (1874a) 'Proposal to apply for anthropological statistics from schools', *Journal of the Anthropological Institute*, vol. 3, pp. 308–11.

Galton, F. (1874b) 'Notes on the Marlborough School statistics', *Journal of the Anthropological Institute*, vol. 4, pp. 130–5.

Galton, F. (1875) 'On the height and weight of boys aged 14 years in town and country Public Schools', *Journal of the Anthropological Institute*, vol. 5, pp. 174–81.

Galton, F. (1878) 'Composite portraits made by combining those of many different persons into a single figure', *Journal of the Anthropological Institute*, vol. 8, pp. 132–48.

Galton, F. (1884a) 'On the Anthropometric Laboratory at the late International Health Exhibition', *Journal of the Anthropological Institute*, vol. 14, pp. 205–19.

Galton, F. (1884b) 'Some results of the Anthropometric Laboratory', *Journal of the Anthropological Institute*, vol. 14, pp. 275–87.

Galton, F. (1884c) *Record of Family Faculties* (London: Macmillan).

Galton, F. (1885) 'Regression towards mediocrity in stature', *Journal of the Anthropological Institute*, vol. 15, pp. 146–263.

Galton, F. (1886) 'Family likeness in stature', *Proceedings of the Royal Society*, vol. 40, pp. 42–73.

Galton, F. (1888a) 'Co-relations and their measurement, chiefly from anthropometric data', *Proceedings of the Royal Society*, vol. 45, pp. 135–45.

Galton, F. (1888b) 'On head growth in students at the University of Cambridge', *Journal of the Anthropological Institute*, vol. 18, pp. 155–6.

Galton, F. (1906) 'Anthropometry at schools', *Journal of Preventive Medicine*, vol. 14, pp. 93–8.

Galton, F. and Mohamed, F. A. (1881) 'An enquiry into the physiognomy of phthisis by the method of composite portraiture', *Guy's Hospital Reports*, third Series, vol. 25, pp. 475–95.

Pearson, K. (1914, 1924 and 1930) *The Life, Letters and Labours of Francis Galton* (3 vols) (Cambridge University Press).

Quetelet, A. (1870) *Anthropometrie, ou mésure des differentes facultés de l'homme* (Brussels: Muquardt).

Quetelet, L. A. J. (1835) *Sur l'homme et le développement de ses facultés. Essai sur physique sociale* (2 vols) (Paris: Bachelier).

Roberts, C. (1874–6) 'The physical development and the proportions of the human body', *St George's Hospital Reports*, vol. 8, pp. 1–48.

Roberts, C. (1878) *A Manual of Anthropometry* (London: Churchill).

Tanner, J. M. (1981) *A History of the Study of Human Growth* (Cambridge University Press).

Tanner, J. M. (1986) 'Use and abuse of growth standards', in: *Human Growth: A Comprehensive Treatise*, F. Falkner and J. M. Tanner (eds), 2nd edn, vol. 3, pp. 95–109 (New York: Plenum).

9 Galton and the Use of Twin Studies

C. G. Nicholas Mascie-Taylor

INTRODUCTION

Galton was fascinated by twins. In this he was not unique, and throughout the ages people have both welcomed twins and feared them. The belief that twins were the possessors of supernatural powers can be traced back to earliest times. Twins sometimes were thought to be the bearers of good fortune or to have second sight. In many of the folk tales the twins appear as gifted heroes, as magicians, as healing gods with miraculous powers such as the ability to forecast the weather, or control it, and the ability to promote or inhibit fertility. Sometimes in these myths, one twin may be good and the other evil. One may symbolise darkness, the other light. One may represent the sun, the other the moon.

There are many examples of twin legends. The native American Indian tribe, the Huron have a story about the twins who they believe founded their tribe. The Huron tale describes the moon goddess (Ataentsic) who fell out of heaven into the primeval waters. Her virgin daughter gave birth to the earth's first children, the twins Iosekha (the White One) and Tawiscara (the Dark One), but died when the Dark One, refusing to be born in the usual manner, burst forth from his mother's armpit. Later the twins quarrelled and Iosekha killed his evil brother with a staghorn. Iosekha then bestowed many gifts on mankind before returning to the sky as the Sun.

The twin hero legend appears in cultures and religions in every part of the world – Castor and Pollux, Romulus and Remus and Esau and Jacob, to name but three. In more modern times twins have appeared on figureheads, in art, in advertising and as subjects for photographers.

THE DISCOVERY OF THE TWIN METHOD

Galton is credited with the use of twins in scientific enquiry and he first mentioned them in his book *English Men of Science*. In it he wrote:

119

There are twins of the same sex so alike in body and mind that not even their own mothers can distinguish them. Their features, voice, and expressions are similar: they see things in the same light, and their ideas follow the same laws of association. This close resemblance necessarily gives way under the gradually accumulated influences of differences of nurture, but it often lasts until manhood (Galton, 1874).

A year later his article 'The History of Twins as a criterion of the relative powers of nature and nurture' was published. It was reprinted the same year with slight revision in the *Journal of the Anthropological Institute* (Galton, 1875). In the article Galton reported on the life histories of two groups of twins, one group consisted of pairs of twins who were similar at birth (35 twin pairs), the other group was dissimilar at birth (20 twin pairs). The similar pairs are probably identical twins and Galton found some changes as well as some consistency in their life histories. Galton's studies on the dissimilar pairs revealed not a single case where an originally dissimilar character 'became more assimilated through identity of nurture'.

Galton was concerned with determining the importance of the environment either in making dissimilar twins more alike or in making similar twins more different. His studies were based on a misconception of the phenomenon of twinning. He thought the same sexed dissimilar twins were identical twins with the dissimilarity arising if the division of the ovum was delayed beyond the point at which differentiation took place.

Another misconception, though not of Galton's making, is the popularly held view that Galton invented or discovered the 'twin method'. This method consists of the comparison of identical and fraternal twins reared together and it is a method widely used in human genetics, particularly in human behavioural genetics, to determine the extent of genetic influences on a trait or character. Galton did not propose the comparison of identical and fraternal twins that is the essence of the twin method. Thus although Galton deserves to be called the father of human behavioural genetics for many reasons, discovering the twin method is not one of them (Rende, Plomin and Vandenberg, 1990).

Who did discover the twin method? Rende, Plomin and Vandenberg (1990) found that the first descriptions of the twin method appeared nearly fifty years after Galton's paper. In 1924 Hermann Siemens, a German dermatologist, published a book which described how the twin method could be used to determine hereditary influence on features such as skin disorders (Siemens, 1924). In the same year Curtis Merriman, an assistant professor of education at the University of Wisconsin, had published an article in *Psychological Monographs* which described the method (Merriman, 1924). However his own twin study only examined the IQs of identical twins and it was not until 1928 that the correlations of monozygotic (MZ, identical) and dizygotic (DZ, fraternal) twins for IQ were first compared (Tallman, 1928; Wingfield, 1928).

THE USE OF TWIN STUDIES

Twinning has been described as 'the most common congenital abnormality', something which twins may find insulting but which at least suggests that twins should be studied not merely as a means of genetic analysis but as a significant and unique portion of the population (Hay, 1985).

Monozygotic twins develop from the splitting of a single fertilised egg whereas dizygotic twins develop simultaneously in the womb from two separate zygotes. Thus monozygotic twins will have identical sets of genes while dizygotic partners will be expected to have only half their genes in common, just like ordinary brothers and sisters.

About one-third of the conceptions that produce twins result in opposite sex fraternal twins, one-third in same sex fraternal twins and one-third in identical twins, although this varies from population to population. These proportions can be calculated using Weinberg's differential method (Weinberg, 1901). In dizygotic twins the proportion of like-sexed twins (male–male and female–female) will be expected to equal the proportion of opposite sex fraternal twins. Thus the number of dizygotic twins can be estimated by doubling the number of unlike-sexed twins. The number of MZ twins is the difference between the numbers of like-sexed and unlike-sexed twins. A worked example is presented below.

Where p is the frequency of males and q of females, for dizygotic twins the frequency of male–male, male–female and female–female twins is given by:

male–male	male–female	female–female
p^2	$2pq$	q^2

Assuming a sex-ratio of 1:1, that is p = 0.5 and q = 0.5 and substituting

| 0.25 | 0.50 | 0.25 |

Thus half of all DZ twins are unlike-sexed twins.

If N = total of maternities, L = number of like-sexed twins and U = number of unlike-sexed twins:

Proportion of DZ twins = $2U/N$

Proportion of DZ twins per 1000 maternities = $(2U/N) \times 1000$

Proportion of MZ twins = $(L - U)/N$

Proportion of MZ twins per 1000 maternities = $([L - U]/N) \times 1000$

For example, in 1973 in Scotland there were 74 500 maternities of which 747 were twin births. The number of like-sexed twins was 516 and unlike sexed twins 231. Therefore:

Frequency of DZ twins = $(2 \times 231)/74\ 500 \times 1000$ = 6.2 per 1000 maternities
Frequency of MZ twins = $(516 - 231)/74\ 500$ = 3.8 per 1000 maternities

This method does not take into account differences in sex ratio (Emery, 1986) and there is a slight excess of like-sexed over unlike-sexed twins (James, 1976).

The frequency of monozygotic twins is very similar throughout the world at between three and four per 1000 births. However dizygotic twinning rates vary quite widely, being lowest in East Asians (between two and eight per 1000 births), intermediate in Europeans (between six and eleven per 1000), and highest in Africans (between ten and forty per 1000). There is a particularly high frequency area in West Africa which includes Mali and Ghana (Chavez and Roberts, 1973: MacGillivray et al, 1975).

The factors determining twinning differ between monozygotic and dizygotic twins. MZ twinning is little influenced by maternal age and parity but the frequency of DZ twins increases significantly with

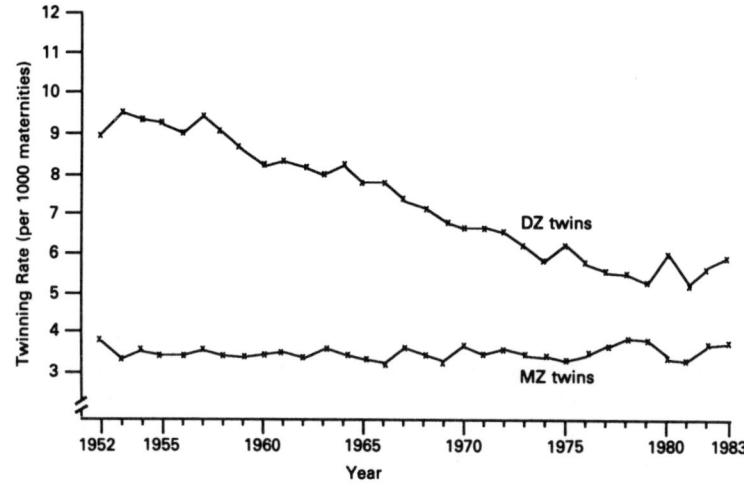

Figure 9.1   DZ and MZ twinning rates (live and still births per 1000
maternities) for England, Scotland and Wales

parity and with maternal age reaching a maximum at 35 to 40 years.
DZ twinning may be partly under genetic control with the mother's
genotype being much more important than the father's. Hereditary
factors seem to be much less important in MZ twinning (Parisi et al.,
1983; Philippe, 1985).

In many developed countries there has been a gradual fall in the
frequency of twin births since the early 1950s. As Figure 9.1 shows,
the decline in England, Wales and Scotland has been due to a
decrease in the frequency of DZ twins, although present evidence
suggests a levelling off and the MZ rate is even rising slightly.

There are a number of reasons for the decline (James, 1972) but
the advent of reliable contraception has meant that older women are
having fewer children. At the same time there are indications of a
higher incidence of MZ twinning among women who conceive within
three months of ceasing to use oral contraceptives due, it is thought,
to delayed implantation of the fertilised ovum (Emery, 1986).

The embryology of twinning has been thoroughly analysed and
discussed by Bulmer (1970). For a single birth the embryo is suspended within a sac formed by two membranes, the amnion and the
chorion. Fraternal twins, which develop from two different zygotes,
usually have a separate amnion and chorion although their placentas
sometimes fuse (Table 9.1). On the other hand monozygotic twins,

Table 9.1　The different types of fetal membranes and their frequencies in dizygotic (DZ) and monozygotic (MZ) twins

Placenta	Chorion	Amnion	DZ (%)	MZ (%)
2	2	2	50	15
1	2	2	50	15
1	1	2	–	70
1	1	1	–	rare

which result from a single zygote, may share a membrane, depending on when the split occurred. If the division of the zygote occurs before implantation (within about five days) then the twins will have separate amnions and chorions just like fraternal twins. Approximately one-third of monozygotic twins have separate amnions and chorions. If the split occurs after implantation the twins develop within the same chorion. When separation occurs five to ten days after fertilisation, as it usually does, there are two amnions. About 4 per cent of monozygotic twins separate after ten days, and these share their amnion as well as chorion. Extremely late separation results in so called 'Siamese' twins where the individuals are partially fused.

It has been suggested that anomalous X inactivation causes some cases of monozygotic twins as well as neural tube defects (Hall, 1986; Burn et al., 1986). James (1991) has recently shown that the later an MZ pair is formed, the more likely it is to be female. This supports the hypothesis that anomalous X-inactivation is involved in the formation of some female MZ twin pairs.

DIAGNOSIS OF ZYGOSITY

The examination of fetal membranes is the time-honoured method of zygosity determination at birth (Emery, 1986). However, as Table 9.1 shows, the diagnosis of zygosity using fetal membranes is not clear cut. In all cases in which there is a single chorion (monochorionic) the twins are unequivocally monozygotic (about 70 per cent of MZ twins). But dichorionic twins can be either MZ or DZ although they are more likely to be dizygotic. Thus an individual dichorionic twin pair of like sex cannot be unequivocally diagnosed.

Furthermore information on fetal membranes may not be available and therefore the 'similarity method' has to be used. In the early days of twin studies, zygosity was based on physical traits such as hair and eye colour. Later blood groups and other genetic markers were used and by 1955 the analysis of blood groups began to be accepted as the best method (Smith and Penrose, 1955).

In 1959 Husen compared the use of physical and blood analysis. He was able to use only four blood markers, of which the ABO blood group system was one. On the basis of the ABO system Husen was able to diagnose accurately 60 per cent of MZ twins since about two-thirds of DZ twins can also be expected to have the same variant by chance alone. The probability that fraternal twins will have the same variant for two blood group systems is much less, so that using ABO and MNS the diagnosis improved to 72 per cent. With all four blood group systems the accuracy was 90 per cent.

Husen also found that physical features provided good precision; 92 per cent of MZ twins had exactly the same eye colour compared with 44 per cent of DZ twins. For hair colour the percentages were 91 per cent and 38 per cent, for hairline pattern, 96 per cent and 16 per cent and for complexion 96 per cent and 55 per cent (MZ and DZ respectively). Even a subjective rating of similarity predicted well. As a general rule, identical twins tend to be misclassified as fraternal twins more often than the opposite (Plomin, DeFries and McClearn, 1990). However just asking twins whether they are identical or fraternal is not a very accurate method (Husen, 1959; Carter-Saltzman and Scarr, 1977). Work by Cohen et al. (1973, 1975) confirmed the accuracy of physical traits in diagnosing zygosity and Table 9.2 presents some of the results of Cohen et al.

Emery (1986) uses data on six blood group systems, secretor status, Haptoglobin type and dermatoglyphics to determine the twin status. In this example (Table 9.3) the parental phenotypes are also known. Only two of the blood groups (ABO and Rh), haptoglobin type and dermatoglyphics provide informative data. The prior probabilities and conditional probabilities are multiplied together to give a joint probability of either dizygosity or monozygosity. The probabilities for the dermatoglyphics were determined from tables (Smith and Penrose, 1955; Smith et al., 1961) which showed that a difference in total ridge count of only four occurs in 4 per cent of like-sexed DZ twins and 27 per cent of MZ twins. A difference in *atd* angles of only 2° occurs in about 9 per cent of DZ twins and 18 per cent of MZ twins.

The results show that there is a 99 per cent probability that the twins are monozygous. If no information was available on the parents of the twins then the probabilities of obtaining the various observed

Table 9.2   Physical similarity and twin zygosity (MZ, monozygous, and DZ, dizygous)

Questions to mothers	Percentage of twins 'exactly similar' or 'yes' responses by mother	
	MZ	DZ
Is it hard for strangers to tell them apart?	100	8
Eye colour?	100	30
Hair colour?	100	10
Facial appearance?	49	0
Complexion?	99	14
Weight?	46	6
Height?	56	13
Do they look alike as two peas in a pod?	48	0
Does either mother or father ever confuse them?	79	1
Are they sometimes confused by other people in the family?	93	1
Number of twin pairs	181	84

traits in the twins would have to have been based on gene frequencies in the general population. The calculations are tedious (Emery, 1986) but tables of relative probabilities of dizygosity are available (Smith and Penrose, 1955; Smith et al., 1961) for blood groups and other traits based on their population frequencies in the UK.

Lykken (1978) has provided a comprehensive review of the diagnosis of zygosity in twins and has discussed the efficiency of a phenotype in discriminating between MZ and DZ twins. His calculations based on the gene frequencies of Minnesota whites indicated that the ABO locus was not a very good discriminator. Its total efficiency is 0.606 (the smaller the value the more efficient is the discrimination) while the MNS blood group was better with an efficiency of 0.438. Multiplying the efficiencies of eight blood groups (ABO, MNS, Rh, Lewis, Kell, Duffy, Kidd and P) yielded a combined efficiency of 0.0018, that is a random pair of DZ twins will by chance be concordant with respect to all eight of these blood groups eighteen times per 1000. Therefore these eight systems alone will discriminate MZ from DZ twins with less than 2 per cent error on average.

When the information on the eight blood groups is combined with

Table 9.3 Calculation of probability of monozygosity

	Father	Mother	Twin A	Twin B
Blood Groups	O	A_1O	A_1	A_1
	R_1r	R_1R_1	R_1R_1	R_1R_1
	MsMs	NsNs	MsNs	MsNs
	P pos	P pos	P pos	P pos
	Lu(a–)	Lu(a–)	Lu(a–)	Lu(a–)
	Kell neg	Kell neg	Kell neg	Kell neg
	Fy(a+)	Fy(a+)	Fy(a+)	Fy(a+)
Secretor status	secretor	secretor	secretor	secretor
Haptoglobin type	1–1	2–1	2–1	2–1
Dermatoglyphics				
Total ridge count	–	–	161	165
Sum of *atd* angles	–	–	86°	88°

Character	PDZ	PMZ
Prior probabilities	0.70	0.30
Conditional probabilities		
Sex	0.50	1.00
Blood Groups		
ABO	0.50	1.00
Rh	0.50	1.00
Haptoglobin type	0.50	1.00
Dermatoglyphics		
Difference in total ridge count	0.04	0.27
Difference in atd angles	0.09	0.18
Joint Probability	0.0001575 (JPDZ)	0.01458 (JPMZ)

$$\text{Probability of dizygosity} = \frac{JPDZ}{(JPDZ + JPMZ)} = \frac{0.0001575}{0.0147375} = 0.0107$$

$$\text{Probability of monozygosity} = \frac{JPMZ}{(JPDZ + JPMZ)} = \frac{0.01458}{0.0147375} = 0.9893$$

ten serum protein and red blood cell enzymes then the eighteen serological markers should misclassify dizygotic twins as monozygotic fewer than two times per 1000. With three additional anthropometric variables, ponderal and cephalic indices and total ridge count Lykken showed that there is only a three in 10 000 chance of misclassification.

The above methods will probably cease to be used as researchers make use of the new molecular genetic methods, in particular the use of DNA 'fingerprinting'. DNA for each person yields a distinct DNA 'fingerprint' of bands with different numbers of tandem repeats (Jeffreys, Wilson and Thein, 1985). The pattern of bands is similar for genetically related individuals, but only identical twins have exactly the same DNA fingerprints. This method provides an excellent test of zygosity without the need for analysis of many individual genetic markers (Plomin, DeFries and McClearn, 1990).

THE USE OF TWINS IN GENETIC ANALYSIS

The twin method was specifically designed to distinguish the influence of genetic factors in given traits. The idea behind the twin method is that monozygotic twins will be genetically identical but may differ from each other due to the environment. Dizygotic twins are only as genetically alike as any brothers or sisters who have half their genes in common, although they do share the same intrauterine environment.

The most commonly employed techniques are the study of concordance rates for discontinuous characters such as disease states, and of variance and correlations for continuous characters such as height, serum lipoproteins and IQ.

Concordance Rates

There are several methods of determining concordance rates. The most common method is the pairwise concordance which is defined as the proportion of affected twin pairs in which both members are affected (concordance rate = $C/C + D$, where C = total number of concordant pairs and D = number of discordant pairs). In Gottesman and Shields' (1972) schizophrenia study, both members of the pair were schizophrenic in eleven of twenty-two pairs of identical twins, so the pairwise concordance is 50 per cent. In the same study, out of thirty-three dizygotic twins three were concordant and the rate is 9 per cent. Other examples of pairwise concordance rates are shown in Table 9.4.

The other main method considers index cases rather than pairs. This is the proband concordance method which is defined as the proportion of affected individuals among the co-twins of previously

Table 9.4 Twin concordance for schizophrenia and common medical disorders

Disorder	Concordance	
	Identical	Fraternal
Schizophrenia	30.9	6.5
Diabetes mellitus	18.8	7.9
Ulcers	23.8	14.8
Pulmonary disease	11.8	8.2
Hypertension	25.9	10.8
Ischemic heart disease	29.1	18.3

ascertained index cases. When both twins are affected and have been independently ascertained, the twin pair is in effect counted twice. Mathematically the proband concordance rate $= C + C'/C + D + C'$, where C = total number of concordant pairs, D = number of discordant pairs and C' = the number of concordant pairs ascertained independently through both affected twins.

In the schizophrenia example of the eleven concordant pairs of MZ twins, only four pairs were independently ascertained, and of the three concordant DZ pairs in one pair both twins had been ascertained independently. Thus the proband concordance rate for MZ twins $= 11 + 4/11 + 11 + 4 = 58$ per cent, and for DZ twins $= 3 + 1/3 + 30 + 1 = 12$ per cent.

Although there are merits to both approaches, on balance the proband concordance rate is more appropriate from a sampling viewpoint because the members of a twin pair are identified independently. In addition the proband concordance can be compared to risk figures for other family groups and to the general population rates.

The interpretation put on concordance rates is that for a disorder in which genetic factors are important in the aetiology, the concordance rate for MZ twins will be greater than for DZ twins. For schizophrenia it would appear that co-twins of a schizophrenic identical twin have a thirty-fold greater risk of becoming schizophrenic compared with individuals in the population, while the risk for DZ twins is only six times greater than the general population. Thus genetic factors appear to be significantly and substantially implicated in schizophrenia. However the concordance of 30 per cent for identical twins means that 70 per cent of the time genetically identical individuals are discordant for schizophrenia, which implies that nongenetic factors primarily determine why one person is diagnosed schizophrenic and another is not (Plomin et al., 1990).

Alcoholism is another characteristic which has been extensively examined. Most studies suggest that alcoholism runs in families and about 25 per cent of the male relatives of alcoholics are themselves alcoholics (Cotton, 1979). No twin studies have focussed on alcoholism *per se* because the relationship between alcohol use and abuse is not understood. One twin study of liver cirrhosis is noteworthy because the major cause of this is advanced alcoholism. In a study of nearly 16 000 middle-aged male twins, the concordance for liver cirrhosis was 15 per cent in MZ twins and 5 per cent in DZ twins (Hrubec and Omenn, 1981).

Continuous Characters

Only a brief review of the algebraic explanation of the partitioning of variance for continuous characters is provided here – more detailed accounts can be found in Hay (1985) and Fulker (1979).

Presented below are the main aspects of the partitioning of variance. It can be seen that the method assumes that the common environments of MZ and DZ twins are about the same. The formula $2(r_{MZ} - r_{DZ})$ somewhat overestimates the broad sense heritability because it contains 1.5 times the dominance variation.

For fraternal twins the components of variance are

$$COV_{DZ} = (1/2V_A) + (1/4V_D) + V_{EC(DZ)}$$

where V_A = additive genetic variance, V_D = dominance genetic variance and $V_{EC(DZ)}$ = common environmental variance.

For monozygotic twins the components of variance are

$$COV_{MZ} = V_A + V_D + V_{EC(MZ)}$$

where $V_{EC(MZ)}$ = common environmental variance.

Thus the difference between the phenotypic covariances of identical and fraternal twins is

$$COV_{MZ} - COV_{DZ} = (1/2V_A) + (3/4V_D)$$

Table 9.5   Correlations for identical and fraternal twins for a number of characters (r = correlation coefficient)

Character	r_{MZ}	r_{DZ}	Heritability
Height	0.93	0.48	0.90
Weight	0.91	0.58	0.66
Total ridge count	0.96	0.49	0.94
IQ	0.86	0.53	0.66
Extraversion	0.51	0.21	0.60
Neuroticism	0.50	0.23	0.54

assuming that the common environments are about the same for the two types of twins

Many researchers prefer to use correlations. The equations then become

$$r_{DZ} = ([1/2V_A] + [1/4V_D] + [V_{EC(DZ)}])/V_P$$

where V_P is the total phenotypic variance and r the intrapair correlation coefficient.

$$r_{MZ} = (V_A + V_D + V_{EC(MZ)})/V_P$$

The difference between the two expressions is

$$r_{MZ} - r_{DZ} = ([1/2V_A] + [3/4V_D])/V_P$$

Doubling this difference gives

$$2(r_{MZ} - r_{DZ}) = (V_A + [3/2V_D])/V_P$$

which is close to the broad sense heritability.

Table 9.5 provides an overview of MZ and DZ correlations for a number of physical and behavioural traits and the broad sense heritability based on the above formula. Heritability estimates have also been obtained for a wide variety of anthropometric characters including head length and breadth, total face height, chest breadth, fat measurements and somatotypes. Figure 9.2 shows the somatotypes obtained from a small Prague study of fourteen pairs of identical twins and ten pairs of fraternal twins. The average difference between components for the identical twins was 0.21 for endomorphy, 0.25 for

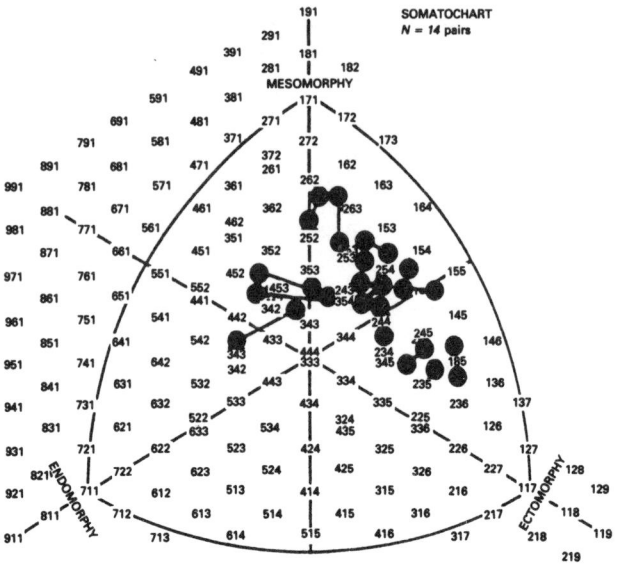

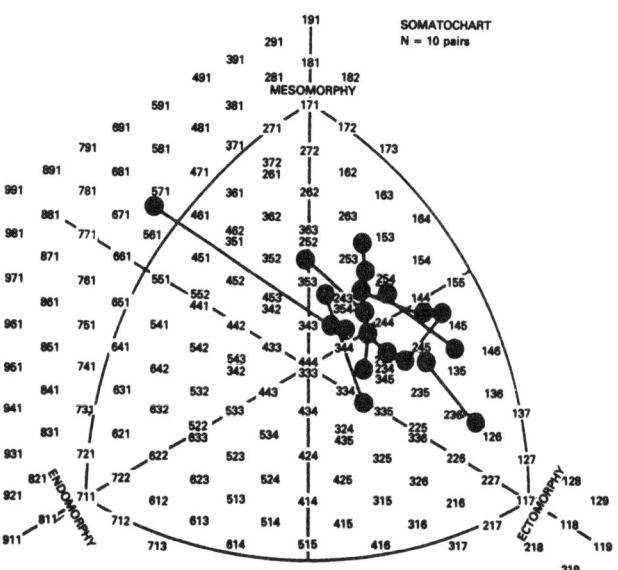

Figure 9.2   (a) Somatotypes of Prague twins: (a) monozygotic twins; (b) dizygotic twins

Source: Redrawn from R. Kovář. 'Somatotype of twins', *Acta Universitatis Carolinae, Gymnica*, vol. 13 (1977), pp. 49–59.

Table 9.6   Correlations between identical twins on behavioural and
environmental measures

	Correlation	Number of pairs
Cognitive	−0.06	276
Personality	0.06	451
Vocational interests	0.01	276
Interpersonal relationships	0.05	276

mesomorphy and 0.28 for ectomorphy. For fraternal twins the differences were 0.75 for endomorphy, 0.90 for mesomorphy and 1.1 for
ectomorphy. The heritability values were 0.69, 0.88 and 0.87 for
endomorphy, mesomorphy and ectomorphy. The heritability of skin
colour has also been determined and values range from 0.37 to 0.83
depending on the site and wavelength used to measure skin relec-
tance (Clark et al., 1981).

The formula $2(r_{MZ} - r_{DZ})$ is only one of a large number that have
been used by researchers studying twins, and no formula allows for
all possibilities of genetic and environmental components and their
interactions. This formula assumes that the common environments
are similar and it also does not allow for epistatic interactions, the
effects of assortative mating, genotype-environment interaction and
correlation. It is perhaps unsurprising that the use of this rather
'crude' formula has been criticised by some workers (for example
Rose, Kamin and Lewontin, 1984) especially when the genetic component of IQ is being discussed!

One of the main criticisms of the twin method is the assumption of
equal environments of MZ and DZ twins and there is a generally held
belief that the environmental experiences of MZ twins are much
more similar than those of DZ twins (Rose et al., 1984). Using three
approaches – (1) studying twins who were misclassified (2) measuring
aspects of the environment of MZ and DZ twins and (3) relating
differences in the environment of the two types of twin to differences
in their behaviour – there is little empirical evidence to support the
unequal environments assumption with regard to behavioural traits.

In the last approach (c) above, Loehlin and Nichols (1976) correlated the behavioural differences with a composite measure of the
environmental difference. The results are presented in Table 9.6

A positive correlation indicates that identical twins exposed to
more similar environments are more similar in behaviour. The low
correlations provide strong support for the equal environments
assumption of the twin method. Greater physical similarity of ident-

Table 9.7 Modelling twin correlations (r = correlation coefficient,
G = genetic effect and E = environmental effect)

Twin type	r	$1-r$
MZ	G	E
DZ	$1/2G$	$(1/2G) + E$

Prediction: $r_{MZ} = 2r_{DZ}$

Table 9.8 Patterns of twin correlations (r = coefficient correlation)

Trait	r_{MZ}	r_{DZ}
Neuroticism	0.50	0.25
Conservatism	0.60	0.40
EEG	0.80	0.00
Lie scale	0.55	0.55

ical twins does not correlate with behavioural measures either
(Matheny, Wilson and Dolan, 1976).

Eaves (1982) has suggested a move away from the formula-
dominated approach in studying twins. He proposes a different
strategy in the analysis of twin data which involves the construction of
models. For example he cites a simple model which would include the
following assumptions: gene action is additive, there is no sex linkage
or sex limitation, no family environmental effects, mating is random,
genes and environment do not interact and genes and environment
are independent. In this case the model would predict that the MZ
twin correlation would be twice the DZ correlation (Table 9.7).

The data in Table 9.8 provide typical patterns of twin correlations.
These are not real data but they do reflect what has been reported in
the literature. For neuroticism the correlation for MZ twins is indeed
twice that for DZ twins and the data agree with the prediction of the
model. Thus we have no reason to modify the model by including any
of the additional effects that were omitted from it.

However for none of the other variables can it be said that the
model fits since the DZ correlation is certainly not half the MZ
correlation. It becomes necessary to look elsewhere for an interpretation of the data. Consider, for example, what would happen if the

Table 9.9   Modified model including between family effects
(r = coefficient correlation, G = genetic effect, B = between family effects,
E = environmental effect)

Twin type	r	$1-r$
MZ	$G + B$	E
DZ	$(1/2G) + B$	$(1/2G) + E$

Prediction: $r_{MZ} < 2r_{DZ}$

Table 9.10   Parameter estimates (G, E and B as in Table 9.9)

Trait	G	E	B	Comments
Neuroticism	0.50	0.50	0.00	Model fits, no family environment
Conservatism	0.40	0.40	0.20	Model fits, significant B
Lie Scale	0.00	0.45	0.55	No genetic effects, large family environment
EEG	1.60	0.20	–0.80	Parameters inconsistent, model fails

environmental effects on twins affected the twins jointly as well as separately. This is not an unreasonable assumption since both MZ and DZ twins live in the same house and share the same parents. The similarity of both MZ and DZ twins is inflated by such effects which are called B for between family effects.

The modified model is shown in Table 9.9. Now it can be seen that the ratio of MZ to DZ correlation is less than two which is exactly what is found for both 'conservatism' and the 'lie' scale. Thus these data seem to fit a model which includes the effects of family environment.

Table 9.10 presents a simple analysis of the twin correlations which includes the family environment. It can be seen for EEG measurement that the observed correlations are not consistent with any known mechanism of genetic transmission.

The examples presented here are oversimplified but they do show the advantage of model building in allowing the comparison of environmental and genetic hypotheses. The models can become more complex to allow for genotype–environment interactions, assortative mating and so on.

Table 9.11   The intraclass correlations between monozygotic and dizygotic
twins (*r* = correlation coefficient)

Study	Number of twin pairs	r
Newman, Freeman and Holzinger		
(1937)	19	0.67
Burt (1966)	53	0.88
Shields (1962)	38	0.78
Juel-Neilsen (1965)	12	0.68
Total	122 weighted average:	0.82

THE SPECIAL CASE OF TWINS REARED APART

Jacob and Esau, like Mark and Carol Thatcher, Castor and Pollux
and Romulus and Remus, were raised together, whether by their
natural parents, the gods or a she-wolf! Twins reared apart (MZA or
DZA) provide the simplest and most powerful method for disen-
tangling the influence of environmental and genetic factors on human
characteristics. The power of the MZA design is that for twins reared
apart from early infancy and randomly placed for adoption, the
shared environmental component is negligible, so the genetic component can be directly estimated from the MZA correlation (Bouchard
et al., 1990).

Until recently the literature on separated monozygotic twins was
slight, with only 122 such twins examined from four published studies. Such a modest total is perhaps not surprising given the rarity of
twins reared apart. Details of the four studies are presented in Table
9.11.

In the mid-1970s it became apparent that many anomalies existed
in Burt's data. The question of how much data Burt actually collected
and how much he fabricated remains obscure. The evidence has been
carefully sifted by Hearnshaw (1979) whose case appears overwhelm-
ing, although recently the debate has been reopened (Fletcher, 1987;
Kline, 1987; Joynson, 1989).

The uncertainty of Burt's data requires their exclusion and there
remain sixty-nine MZA twin pairs with a weighted correlation of
0.74. However all three of these studies have been criticised by
Kamin (1974), an ardent antihereditarian. The main criticism is that
separated twins are never assigned at random to the whole range of

different environments. In some cases parents of twins have been reluctant to rear both twins and instead have opted to keep one themselves and to place the other with relatives or acquaintances of similar socio-economic level as themselves. In the view of critics, therefore, the high correlations between twins in different homes arise largely, or wholly, because of selective placement in similar homes. What the critics fail to address is why the similarity of environment is greater for MZ in different homes than it is for DZ reared together in the same home.

The extent to which Kamin went to deride the genetic hypothesis knew no bounds. For instance twenty-seven of the forty pairs of twins studied by Shields were raised by relatives. Kamin calculates the inter-twin correlation for those reared with relatives was 0.83, whereas for the thirteen pairs with unrelated foster parents the correlation was 0.51. The latter correlation is similar to that observed for DZ twins reared together.

Fulker (1975) pointed out that first cousins are likewise reared by relatives and often have childhood contacts, yet their intercorrelation is 0.26 not 0.83. In addition the group with the more dissimilar environments, which gave the correlation of 0.51, happened to include three highly bizarre and abnormal pairs, and omitting these brings the correlation for the ten remaining pairs up to the same level as for the twenty-seven reared with relatives!

In 1990 the results of one more IQ study of twins reared apart has been reported. This is the Minnesota Study under the direction of Professor Tom Bouchard (Bouchard et al., 1990). Bouchard and his colleagues have studied more than 100 sets of reared apart twins or triplets from the USA and the UK over the past eleven years, although the reported results only concern fifty-six MZAs.

Zygosity in the Minnesota study was based on serological, fingerprint and anthropometric measurements and the probability of misclassification was less than 0.001. Each participant completed about fifty hours of medical and psychological assessment and separate examiners administered the IQ tests. All the twins spent their formative years apart. The total contact time of the twins was quite variable, with a mean of 112.5 weeks (Table 9.12).

The interclass correlations for MZAs, a comparison group of MZTs and retest stability coefficients are shown in Table 9.13 together with the MZA/MZT ratio. For finger ridge count and height there is only a small difference between correlations of MZA and MZT. For the WAIS IQ test the results suggest that about 70 per cent

Table 9.12   The Minnesota separated twins study

Statistics	Age	Time together prior to separation (months)	Time apart to first reunion (years)	Total contact time (weeks)	IQ
Mean	41.0	5.1	30.0	112.5	108.1
SD	12.0	8.5	14.3	230.7	10.8
Range	19.0–68.0	0–48.7	0.5–64.7	1.0–1233.0	79.0–133.0

Source: Bouchard et al., 'Sources of Human Psychological Differences: the Minnesota study of twins reared apart', *Science*, vol. 250 (1990), pp. 223–8.

Table 9.13   Interclass correlations (*r*), sample sizes (*N*) and MZA/MZT ratio of monozygotic twins reared apart and monozygotic twins reared together

Variable	r MZA	N	r MZT	N	Relia-bility	rMZA/rMZT
Total ridge count	0.97	54	0.96	274	0.99	1.01
Height	0.86	56	0.93	274	0.98	0.925
Weight	0.73	56	0.83	274	NA	0.880
WAIS Full scale	0.69	48	0.88	40	0.90	0.784
WAIS Verbal	0.64	48	0.88	40	0.84	0.727
WAIS Spatial processing	0.71	48	0.79	40	0.86	0.899

of the observed variation in IQ can be attributed to genetic variation. This figure is very close to the value of 66 per cent obtained from MZT and DZT comparison.

In addition to the scientific data Bouchard's study has uncovered astonishing similarities between MZAs. Among the first pairs to be studied were the 'Jim Twins'. They were placed for adoption immediately after birth and were separated at the age of four weeks, when one was adopted by the Springers and the other by the Lewises. They were reunited thirty-nine years later. It emerged that both men were divorced from women named Linda and married to women named Betty. Springer named his first son James Allan and Lewis named his first son James Alan. Each has an adopted brother named Larry; each a pet dog named Toy; each drove a Chevrolet from Ohio to the same vacation area – a three-block-long beach area in Florida.

Other separated identical twins also show similar types of resemblances but the cause(s) is still unknown.

REPRESENTATIVENESS OF TWINS

One of the questions which is regularly asked of twins is whether they are representative of the population as a whole. Does a genetic analysis based solely on twins apply to the population as a whole? One way in which twins are different from the non-twin population is that twins are on average three to four weeks premature compared to singletons. In addition they are 30 per cent lighter and 17 per cent shorter than singletons on average but these differences disappear by school age.

There is also some evidence from the 1970s that twins do not do as well as single-born children on a verbal reasoning test and this difference was consistent over a range of family sizes and maternal ages (Record, McKeown and Edwards, 1970). The average score of twins was 95 compared with the singleton mean of 100. Twins in which the co-twin had died either at or shortly after birth had an average of 99, suggesting that the causes of the decrement must be postnatal. More recent studies suggest that most of the twin deficit is recovered early in the school years (Wilson, 1983).

GALTON AND DEVELOPMENT

In his original twin study in 1875 Galton addressed the question of whether the similarity or dissimilarity of twin pairs changed during development. Subsequent studies by Thorndike in 1905 and Merriman in 1924 concerned the same issue. In all three studies no evidence was found for different twin correlations in younger and older age groups, suggesting that the twin similarity remains relatively constant from childhood through to adolescence.

However twin correlations for IQ do change with age and it is not until about seven years of age that they reach levels comparable to those of adolescent and adult twins. Heritabilities are low up to four years of age, of the order of 0.1, and then rise rapidly up to 0.5 by the time the child is aged seven or eight. This pattern is very different to that observed for height, where the heritability is moderately high by one year of age (about 0.4). It then increases to about 0.7 by two years of age, remains constant until four years of age and then increases to adult levels.

CONCLUSIONS

The goal of human quantitative genetics is to infer accurately the causes of phenotypic variation in defined populations (Nance, 1978). There are a limited number of genetic relationships that are available for analysis and twins have long been used in genetics as an adjunct to studies of nuclear families and the adoption paradigm. Even so, twins studies have been subjected to intensive criticism on biological, methodological and epidemiological grounds. Most criticisms of twin methods have attempted to demonstrate that they lead to overestimation of the role of heredity. Yet a closer examination of many of these so-called biases suggests that they do not necessarily operate in such a manner. The recent work on separated identical twins suggests that twins studies will continue to remain a visible and powerful tool in the armamentarium of human genetics.

There are other twin study designs besides those mentioned here. They include studies of families of identical twins, identical co-twin control studies, and studies of genetic and phenotypic similarity within pairs of fraternal twins. A review of these designs can be found in Plomin et al. (1990).

It can be seen that Galton began a tradition of behavioural research with twins that has continued uninterrupted to this day. The methodology has become more intricate but the fact remains that Galton was ahead of his time in seeing the importance of twin studies and he drew strong conclusions from his work with twins. He wrote:

There is no escape from the conclusion that nature prevails enormously over nurture when the differences of nurture do not exceed what is commonly found among persons of the same rank of society and in the same country. My only fear is that my evidence seems to prove too much and may be discredited on that account, as it seems contrary to all experience that nurture should go for so little. (Galton, 1875).

References

Bouchard, T. J., D. T. Lykken, M. McGue, N. L. Segal and A. Tellegen, 'Sources of Human Psychological Differences: the Minnesota study of twins reared apart', *Science*, vol. 250, pp. 223–8.

Bulmer, M. G. (1970) *The Biology of Twinning in Man* (Oxford: Clarendon Press).

Burn, J., S. Povey, Y. Boyd, E. A. Munro, L. West, K. Harper and D. Thomas (1986) 'Duchenne muscular dystrophy in one of monozygotic twin girls', *Journal of Medical Genetics*, vol. 23, pp. 494–500.

Burt, C. L. (1966) 'The genetic determination of differences in intelligence: a study of monozygotic twins reared together and apart', *British Journal of Psychology*, vol. 57, pp. 137–53.

Carter-Saltzman, L. and S. Scarr (1977) 'MZ or DZ? Only your blood grouping knows for sure', *Behavior Genetics*, vol. 7, pp. 273–80.

Chavez, J. B. and D. F. Roberts (1973) 'Twin Studies', in A. Basu, A. K. Ghosh, S. K. Biswas and R. Ghosh (eds), *Physical Anthropology and its Extending Horizons* (Calcutta: Orient Longman).

Clark, P., A. E. Stark, R. J. Walsh, R. Jardine and N. G. Martin (1981) 'A twin study of skin reflectance', *Annals of Human Biology*, vol. 8, pp. 529–41.

Cohen, D. J., E. Dibble, J. M. Grawe and W. Pollin (1973) Separating identical from fraternal twins', *Archives of General Psychiatry*, vol. 29, pp. 465–9.

Cohen, D. J., E. Dibble, J. M. Grawe and W. Pollin (1975) 'Reliably separating identical from fraternal twins', *Archives of General Psychiatry*, vol. 32, pp. 1371–5.

Cotton, N. S. (1979) 'The familial incidence of alcoholism: a review', *Journal of Studies in Alcohol*, vol. 40, pp. 89–116.

Eaves, L. J. (1982) 'The utility of twins', in V. E. Anderson, W. A. Hauner, J. K. Penry and C. F. Sing (eds) *Genetic Basis of Epilepsies* (New York: Raven Press) pp. 249–76.

Emery, A. E. H. (1986) 'Identical twinning and oral contraception', *Biology and Society*, vol. 3, pp. 23–7.

Emery, A. E. H (1986) *Methodology in Medical Genetics*, 2nd edn (London: Churchill).

Fletcher, R. (1987) 'The doubtful case of Sir Cyril Burt', *Social Policy and Administration*, vol. 21, pp. 40–57.

Fulker, D. W. (1975) 'Review of The Science and Politics of IQ by L. J. Kamin', *American Journal of Psychology*, vol. 88, pp. 505–19.

Fulker, D. W. (1979) 'Some implications of biometrical genetical analysis for psychological research', in J. R. Royce and L. P. Mos. Alphen aan den Rijn (eds) *Theoretical Advances in Behavior Genetics* (Netherlands: Sijthoff Noordoff International).

Galton, F. (1874) *English Men of Science: their Nature and Nurture* (London: Macmillan).

Galton, F. (1875) 'The history of twins, as a criterion of the relative powers of nature and nurture', *Fraser's Magazine*, vol. 12, pp. 566–76. Revised version reprinted in *Journal of The Anthropological Institute*, vol. 5, pp. 391–406, 1876.

Gottesman, I. I. and J. Shields (1972) *Schizophrenia and Genetics*: a twin study vantage point (New York: Academic Press).

Hall, J. G. (1986) 'Neural tube defects, sex ratios and X-inactivation', *Lancet*, vol. ii, pp. 1334–5.

Hay, D. A. (1985) *Essentials of Behavior Genetics* (London: Blackwell Scientific Publications).

Hearnshaw, L. S. (1979) *Cyril Burt, Psychologist* (London: Hodder and Stoughton).

Hrubec, Z. and G. S. Omenn (1981) 'Evidence of genetic predisposition to alcohol cirrhosis and psychosis: twin concordances for alcoholism and its biological end points by zygosity among male veterans', *Alcoholism: Clinical and Experimental Research*, vol. 5, pp. 207–15.

Husen, T. (1959) *Psychological Twin Research* (Stockholm: Almqvist and Wiksell).

James, W. H. (1972) 'Secular changes in dizygotic twinning rates', *Journal of Biosocial Science*, vol. 4, pp. 427–34.

James, W. H. (1976) 'The possibility of a flaw underlying Weinberg's differential rule', *Annals of Human Genetics*, vol. 40, pp. 197–9.

James, W. H. (1991) 'A further note on the sex ratio of monoamniotic twins', *Annals of Human Biology*, vol. 18, pp. 471–4.

Jeffreys, A. J., V. Wilson and S. L. Thein (1985) 'Individual-specific "fingerprints" of human DNA', *Nature*, vol. 316, pp. 76–9.

Joynson, R. B. (1989) *The Burt Affair* (London: Routledge/Chapman & Hall).

Juel-Nielsen, N. (1965) 'Individual and environment: a psychiatric–psychological investigation of monozygotic twins reared apart', *Acta Psychiatrica et Neurologica Scandinavica* (Monograph Supplement) vol. 183.

Kamin, L. J. (1974) *The Science and Politics of IQ* (Potomac, Maryland: Lawrence Erlbaum).

Kline, P. (1987) 'Comments on Ronald Fletcher's defence of Cyril Burt', *Social Policy and Administration*, vol. 21, pp. 105–8.

Kovář, R. (1977) 'Somatotype of twins', *Acta Universitatis Carolinae, Gymnica*, vol. 13, pp. 49–59.

Loehlin, J. C. and R. C. Nichols (1976) *Heredity, environment and personality* (Austin: University of Texas Press).

Lykken, D. T. (1978) 'The diagnosis of zygosity in twins', *Behavior Genetics*, vol. 8, pp. 437–73.

MacGillivray, L., P. P. S. Nylander and G. Corney (1975) *Human multiple reproduction* (London: Saunders).

Matheny, A. P., R. S. Wilson and A. B. Dolan (1976) 'Relations between twins' similarity of appearance and behavioral similarity: testing an assumption', *Bahavior Genetics*, vol. 6, pp. 343–52.

Merriman, C. (1924) 'The intellectual resemblances of twins', *Psychological Monographs*, vol. 33, pp. 1–58.

Nance, W. E. (1978) 'The role of twin studies in human quantitative genetics', in W. E. Nance (ed.) *Twin Research* (New York: Alan R. Liss) pp. 73–107.

Newman, H. H., F. N. Freeman and K. J. Holzinger (1937) *Twins: a study of heredity and environment* (University of Chicago Press).

Parisi, P., M. Gatti, G. Prinzi and G. Caperna (1983) 'Familial incidence of twinning', *Nature*, vol. 304, pp. 626–8.

Philippe, P. (1985) 'Genetic epidemiology of twinning: a population based study', *American Journal of Medical Genetics*, vol. 20, pp. 97–104.

Plomin, R., J. C. DeFries and G. E. McClearn (1990) *Behavioral Genetics: a primer*, 2nd edn (New York: Freeman and Company).

Record, R. G., T. McKeown and J. H. Edwards (1970) 'An investigation of the difference in measured intelligence between twins and single births', *Annals of Human Genetics*, vol. 34, pp. 11–20.

Rende, R. D., R. Plomin and S. G. Vandenberg (1990) 'Who discovered the twin method?', *Behavior Genetics*, vol. 20, pp. 277–85.

Rose, S. D., L. J. Kamin and R. C. Lewontin (1984) *Not in our Genes* (London: Penguin Books).

Shields, J. (1962) *Monozygotic Twins* (London: Oxford University Press).

Siemens, H. (1924) *Die Zwillingspathologie* (Berlin: Springer-Verlag).

Smith, S. M. and L. S. Penrose (1955) 'Monozygotic and dizygotic twin analysis', *Annals of Human Genetics*, vol. 19, pp. 273–89.

Smith, S. M., L. S. Penrose and C. A. B. Smith (1961) *Mathematical Tables for Research Workers in Human Genetics* (London: Churchill).

Tallman, G. G. (1928) 'A comparative study of identical and nonidentical twins with respect to intelligence resemblances', *Twenty-seventh yearbook of the National Society for Studies in Education*, pp. 83–6.

Thorndike, E. L. (1905) *Measurement of twins* (New York: Science Press).

Wilson, R. S. (1983) 'The Louisville Twin Study: developmental synchronies in behavior', *Child Development*, vol. 54, pp. 298–316.

Weinberg, W. (1901) 'Beiträge zur Physiologie und Pathologie der Mehrlingsgeburten beim Menschen', *Pflugers Archiv fur die gesamte Physiologie des Menschen und der Tiere*, vol. 88, pp. 346–430.

Wingfield, A. H. (1928) *Twins and Orphans: the inheritance of intelligence* (London: Dent).

10 Galton and the Study of Fingerprints

Gertrud Hauser

On the surface of the palms there are flexion creases that were used by Gypsy Rose Lee and other palmists to tell fortunes. On closer inspection it is clear that there are also minute ridges separated by furrows which form patterns. These anatomical structures were also seen by Galton and countless others before him, by scientists, doctors, lawyers, government officials and artists who depicted, carved, printed, made impressions of them and wrote about them, even as far back as antiquity (Heindl, 1927). But it was only Galton who extracted the ore from these various lodes to forge a new science – that of *dermatoglyphics*.

Maintaining the analogy, the lode was struck, as with so many rich discoveries, almost by chance. Galton noted (Galton, 1891) that his attention was first drawn to finger ridges when preparing a lecture on personal identification that was principally devoted to examining Bertillon's anthropometric method, then newly introduced in France to identify prisoners. For, in preparing that lecture in 1888, he was surprised to find regarding finger marks 'both how much had been done and how much remained to do before establishing their theoretical value and practical utility' (Galton, 1892, p. 2). He resolved to investigate the subject, and his principal contributions may be considered under two headings: the biology of dermatoglyphics, and their applications.

The prerequisite for both was obvious – prints sufficiently clear to allow even small details to be seen. Among the methods he drew together, and with which he experimented, he decided on the printer's ink method with a slab and roller as that which would be used in his own laboratory. This method is still widely used as a fool-proof printing technique today. Among the other techniques with which he experimented, he was one of the pioneers in the use of photography. Interest has recently been awakened in others, which he described as result of new materials and technological advances, for example adhesive tape and computerised photography.

Validation of Galton's position in dermatoglyphic research is done best in the light of the existing knowledge on the subject at his time.

ANATOMY

The first descriptive anatomical studies date back to the seventeenth century. Among these the works of Grew (1684), Bidloo (1685) and Malpighii (1686) are best known. They contain descriptions of sweat pores, epidermal ridges and case reports of their arrangements, some of them including exact drawings of ridge configurations (Figure 10.1). Galton does not refer to them but he cites Purkinje (1823) whose publication represents a milestone in the history of dermatoglyphics in that he was the first to classify systematically the varieties of

Figure 10.1   Mayer's drawings of fingerprints (1788)

Source: J. C. A. Mayer, *Anatomische Kupfertafeln nebst dazugehörigen Erklärungen* (Berlin: J. Decker and H. A. Rottman, 1783–94).

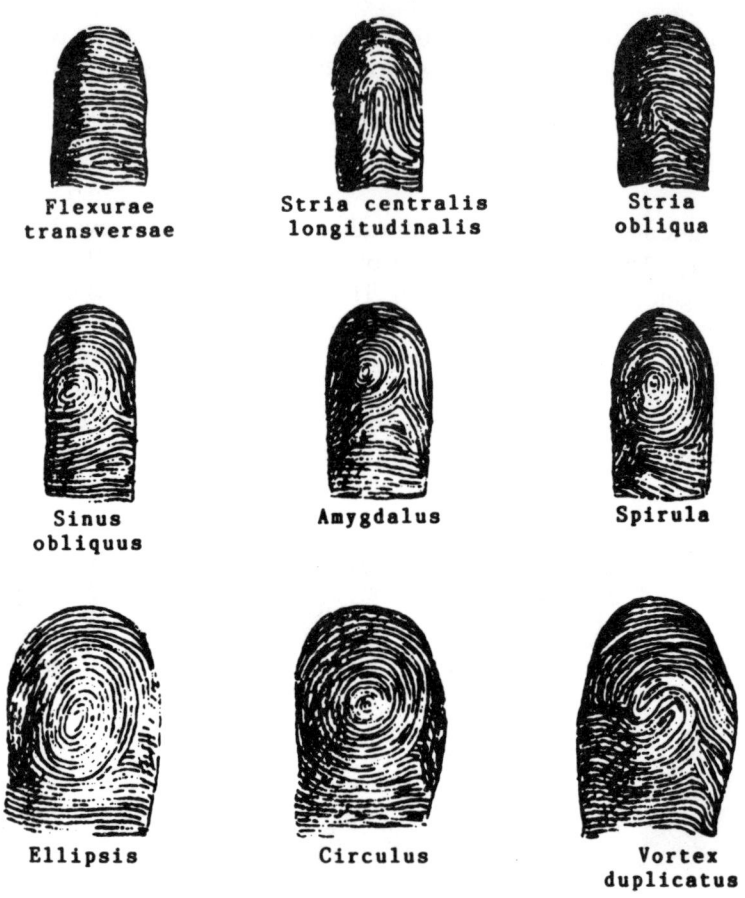

Flexurae   **Stria centralis**   **Stria**
transversae   **longitudinalis**   **obliqua**

Sinus   **Amygdalus**   **Spirula**
obliquus

Ellipsis   **Circulus**   **Vortex**
                                             duplicatus

Source: J. E. Purkinje, *Commentatio de examine physiologico organi visus et systematis cutanei* (Vratislaviae: Typis Universitatis, 1823).

Figure 10.2   The nine fingerprint pattern types

patterns on the fingers. He distinguished 'novem principales formae vallecularum tactui inservientium in extremis manus humanae phalangibus obviae' (nine principal configurations of the valleculae serving the sense of touch on the terminal phalanges of the human hand: Figure 10.2). Purkinje however in no way mentions any other aspect connected with variation of the finger patterns, though he was also the first to point out the importance of 'trianguli' for pattern classification. Purkinje's work was little known as very few copies of it were printed, and these were not widely distributed and therefore not easily accessible.

EMBRYOLOGY

A first description of early fetal development of epidermal ridges was given by Kollmann (1883) who reported formation of dermato-glyphics between the fourth and the sixth fetal month. This has been corrected in the meantime (Heindl, 1927; Bonnevie, 1927, 1929; Schaeuble, 1933; Okajima, 1975; Babler, 1979). The critical stages of dermal ridge differentiation on the fingers range from the tenth to the eleventh week (primary dermal ridges) to the fifteenth and twentieth week (secondary dermal ridges). The importance of such early pre-natal formation of dermatoglyphics however did not escape Galton's attention even though he did not investigate their embryology himself – he reported Kollmann's findings in his book *Finger Prints* (1892) in the chapter 'The Ridges and Their Uses'.

DERMATOGLYPHICS IN POSTNATAL LIFE

There existed but little and by no means sufficient evidence for the three notable characteristics investigated and proved by Galton which make dermatoglyphics important both for academic human biology and for personal identification.

1. Dermatoglyphics do not undergo age changes after their forma-tion.
2. They are not affected by environment after their formation.
3. The detailed structures of individual ridges are extremely variable.

Galton's great merit, besides drawing together information on all these topics, was the study of thousands of prints and relating his observations to other bodily structures. Thus he was the first to mention that effects of aging during childhood and adolescence are limited to increase of ridge size with ongoing growth of the body. Ridges are extremely narrow in the infant and gradually broaden and become higher as the child grows; but there are no changes in their original characteristics of branching, ending or other detail.

There existed more information on permanence of ridge configuration in later life. The most important information came from Faulds, who mentioned persistence in 1880 but without supporting his statement, and Herschel (1880) who had obtained some evidence for it. It also appears that a German anatomist had already made a print of his own left hand in 1856 and repeated it in 1897 at the age of 75 (Figure 10.3) prompted by Galton's publications to prove permanence also for palmar configurations (Welcker, 1898). With respect to environmental effects, Galton mentioned effect of work and effect of injuries. Regarding the former there is a rather amusing quotation: 'They are but faintly developed in the hands of ladies rendered delicate by the continuous use of gloves and lack of manual labour, and in idiots of the lowest type who are incapable of labouring at *all*'. His observations were that the ridges appear stronger in those habitually engaged in moderate manual labour. This we know today is due to moderate thickening of the stratum corneum. He was correct in saying that formation of callosities on the other hand due to hard manual labour may reduce ridge height to such an extent that they are difficult to see and also to print. But his interpretation regarding females and idiots is not totally correct, for there is generally a qualitative sex difference and also a qualitative difference between trisomy 21 patients and normals (Poll, 1934; Abel 1938; Geipel, 1961).

A quite spectacular report of complete ridge regeneration after superficial injury came from Faulds (1905), who reported experiments whereby the surface skin had been shaved off the finger, and prints taken before and after the skin had regrown provided evidence of ridge regeneration.

This however is different with deeper injuries, as mentioned by Galton in 1892: 'A deep clean cut leaves a permanent thin mark across the ridges, sometimes without any accompanying puckering; but there is often a displacement of the ridges on both sides of it . . . other injury that is not a clean incision, leaves a scar with puckerings on all sides, making the ridges at that part indecipherable, even if it does not wholly obliterate them'. It is now well established that when there is injury that is not superficial but extends to a depth of a millimeter and more, causing destruction of the stratum basilare of the epidermis where regeneration occurs, ridges grow distorted and scars result. Galton gave an impressive practical demonstration of this. In 1896 he showed from a print that a graft that had been cut off by a man while cutting cardboard with a sharp knife had been replaced crossways to its original position (Figure 10.4).

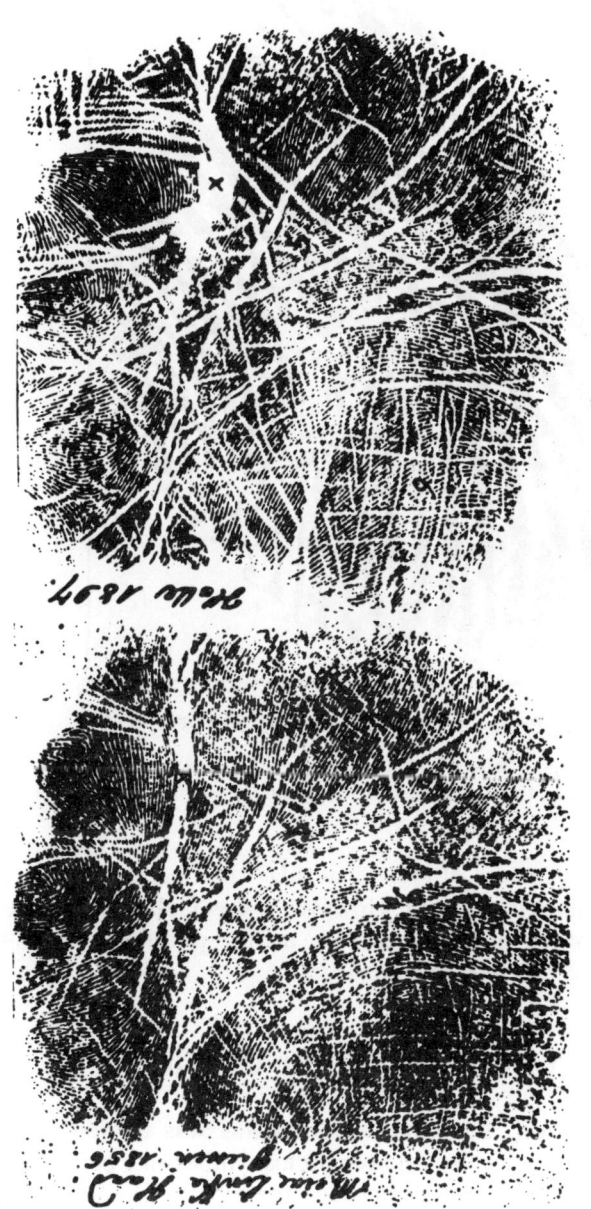

Source: H. Welcker, 'Die Dauerhaftigkeit des Dessins der Riefchen und Fältchen der Hände', *Arch. f. Anthropologie*, vol. 25 (1898), pp. 29–32.

Figure 10.3 Left palm printed by Welker in 1856 (left) and in 1897 (right)

Source: F. Galton, 'Prints of Scars', *Nature*, vol. 53 (1896) p. 295.

Figure 10.4   Enlarged print of a misplaced graft of flesh on the side of a
thumb, thirty years after it was made

The detailed structure of individual ridges had been cited by several of the authors mentioned. However it was Galton who gave a detailed description of the 'numerous minute peculiarities' (1891) which he named minutiae (Figure 10.5).

DESCRIPTION OF PATTERNS ON FINGERTIPS

As has been shown, the basic work on this was carried out by Purkinje (1823). To this Galton brought his synthesising perspective, and he perceived relationships between these patterns and grouped them accordingly. After laborious trials, originating from some sixty standards, he came to differentiate three main classes of fingertip patterns according to the number of triradii present (triradius was the

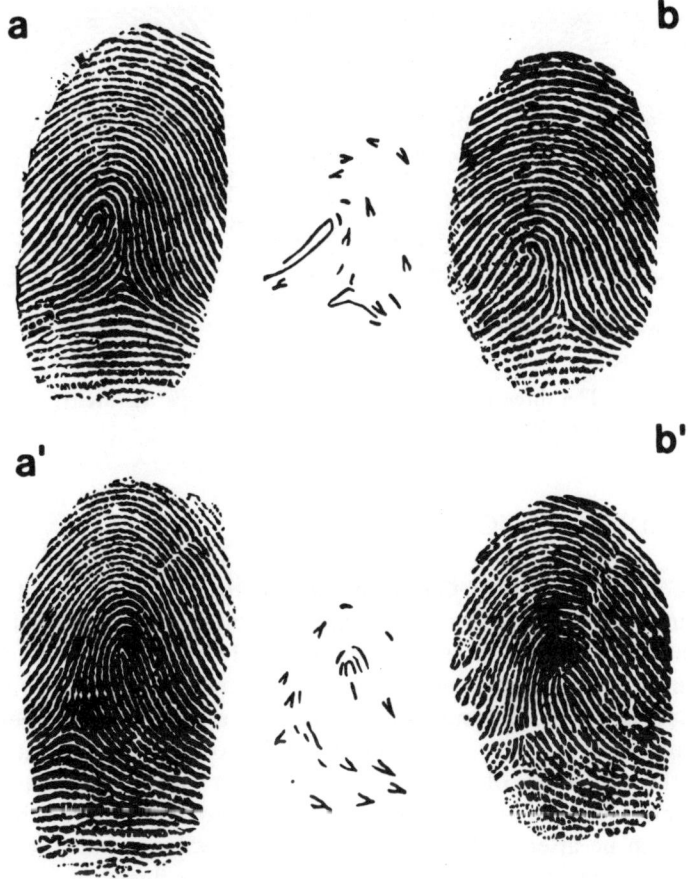

Source: F. Galton, 'Personal Identification and Description', *Journal of the Anthropological Institute*, London, vol. 18 (1888) pp. 177–91.

Figure 10.5   Enlarged impressions of the fore (a) and middle (b) fingertips of the right hand of Sir William Herschel, made in 1860 and in 1888 (a', b'), and positions of 'furrow heads and bifurcations' of the prints made in 1860 (a, b)

name given by him to Purkinje's triangulus). They represent a triradiate structure – the meeting point of three ridge systems (Figure 10.6). His primary pattern, an arch composed of a succession of gently curving ridges, has no triradius. The second class of pattern (the loop) has one triradius, and the third class (the whorl) has two or more triradii. This classification is still the basis of all present work concerned with dermatoglyphics.

Figure 10.6   The three main pattern types on fingertips: A (arch),
L (loop), W (whorl)

INDEXING OF FINGERPRINTS

There is no doubt that Galton pioneered the development of the system of identification by fingerprints. Galton was fully aware of the enormous variety of fingerprint patterns. He also knew that for a system of indexing and recording to be practicable it needed to be simple. He therefore decided to base his system on a small number of principal features rather than a large number of details. The features were the occurrence of an arch (A), loop (L) or whorl (W) on each of the digits arranged in a standard order. The sequence that he chose, but which he afterwards modified (Galton, 1895) was index-middle ring finger on the right hand, index-middle-ring finger on the left hand, thumb-little finger on the right hand and finally thumb-little finger on the left hand. The pattern on each of these fingers was described as A, L or W but that for the index finger was further refined according to whether any loop that occurred was ulnar or radial, which he called outer or inner (Figure 10.5). For example a reading of AWL ALL WW LW would signify an arch on both index fingers; a whorl on the right and a loop on the left middle finger and loops on both ring fingers; a whorl on the right, a loop on the left thumb; and a whorl on both little fingers. Each of these he was able to justify on the basis of his knowledge of pattern occurrence and distribution across the fingers.

This system is essentially a guide to data storage and recovery and once a group of potential matching individuals is extracted by it then the minutiae, scars and so forth can be compared. This simple system of classification was adopted in essence by police forces – in 1901 by Scotland Yard, and on the continent by the police in Budapest and Vienna in 1902 (Heindl, 1927). After one year of use of 'dactyloscopie' in London the success was four times that of five years without dactyloscopie, besides the enormous gain of time (Heindl, 1927).

A most striking illustration of Galton's legacy occurred in Blackburn nearly forty years after his death (Thorwald, 1965). The murderer of a child was identified by taking the fingerprints of the total male population of the town, about 5000 individuals. Unfortunately the material had to be destroyed afterwards to placate public opinion, which considered storage to be an infringement of personal freedom.

HEREDITY

For those of us concerned with the Human Genome Project it is interesting that Galton contemplated an enquiry to determine the smallest biological unit that may be 'hereditarily transmissible'. He was of course not thinking of DNA but regarded the minutiae in fingerprints as suitable for this purpose. This again is a topic that has been developed in recent decades (Ökrös, 1965; Steffens, 1965; Okajima, 1967; Loesch, 1973). Galton's role in laying the foundations of genetics of continuosly varying characters is well known. It is less well known that he was also the true pioneer in studies of inheritance of dermatoglyphics.

Galton himself relates how this came about: 'after examining many prints, the frequency with which some peculiar pattern was found to characterise members of the same family convinced me of the reality of an hereditary tendency'. In order to study such genetic influence he compared the incidence of the three pattern types in pairs of sibs and in pairs of children picked out at random. The observed number of times in which there was correspondence of patterns on the right forefinger between sibs was consistently greater than those of the unrelated. He expanded this method to analyse 53 pattern types instead of three with a similar result.

It was natural that Galton should have included some data on twins in his analysis. He even mentioned that 'the signs of heredity between brothers and sisters ought to be especially apparent between twins of the same sex, who are physiologically related in a peculiar degree and are sometimes extraordinarily alike'. But being unaware of the importance of dizygous and monozygous differentiation, his results could only demonstrate a strong tendency towards resemblance in the fingerprint patterns in twins.

One of Galton's tentative conclusions that has been substantiated in more recent work was his finding of greater maternal influence than paternal (Knussman, 1973; Reed et al., 1978). This is compatible with modern knowledge of the quantitative roles of the X and Y chromosomes and of cytoplasmic transmission (Penrose, 1967).

VARIATIONS WITHIN AND AMONG POPULATIONS

Galton's attempts to establish the varying similarities between digits foreshadowed the interdigital correlation studies. These in turn led to the application of multivariate concepts such as factor analysis (Knussmann, 1967; Jantz and Owsley, 1977; Reed et al., 1978) and principal components analysis (Jantz, 1974; Roberts and Coope, 1975), and to modern concepts of control of development of digital dermatoglyphics on the extremity as a whole as envisaged in dermatoglyphic field theory (Roberts and Coope, 1979).

Much of what Galton pointed out regarding the importance of dermatoglyphics in biology is of modern relevance. Panmixia exists in respect to dermatoglyphic patterns for they cannot exercise the slightest influence on marriage selection, and therefore are of particular interest as models for quantitative inheritance.

Galton was interested in the possibility of dermatoglyphic differences among populations. He carried out a small survey but concluded that these were negligible. Kollmann (1883) before him had left this question open. Today, with a greater amount of information available, we know that populations may be characterised by different frequencies of dermatoglyphic traits but that no pattern is specific to a given population. These frequency differences have been used in local studies in genetic differentiation (Hauser, 1986). It is interesting from the point of view of population psychology that the Chinese, whose frequency of whorls is among the highest in the world (Chamla, 1962–3; Plato, 1978), called a whorl a snail and a loop a sieve (Figure 10.7) and interpreted them as signs of good and bad luck (Heindl, 1927). According to Heindl, and as I have myself seen on statues, Buddha is frequently depicted as having a whorl on every finger. This may be considered as a purposeful attribute to the incarnation of perfect state.

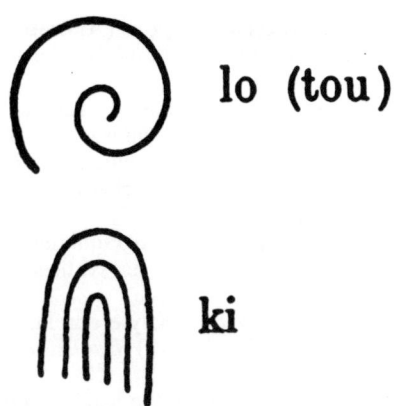

lo (tou)

ki

Figure 10.7   Chinese drawing of a whorl (snail) and a loop (sieve)

CONCLUSION

Summarising Galton's contribution to the study of fingerprints, he appears as the principal protagonist for their study and their application, partly because of the publicity that he gave them, partly because of the way he drew together different aspects, and partly because of his efforts to back his statements by accumulating a mass of data.

It was by bringing together the different kinds of existing but dispersed knowledge, and by his thoughts and his own work, that Galton laid down the tracks for future exploitation. He was the signalman who switched the points so that the train of progress was directed along more modern lines.

References

Abel, W. (1938) 'Kritische Studien über die Entwicklung der Papillarmuster auf den Fingerbeeren', *Z.Menschl.Vererb.u.Konstit.Lehre*, vol. 21, pp. 529–97.

Babler, W. J. (1979) 'Quantitative differences in morphogenesis of human epidermal ridges', *Birth Defects*, orig. article series vol. 15, part 6, pp. 199–208.

Babler, W. J. (1987) 'Prenatal development of dermatoglyphic digital patterns: associations with epidermal ridge, volar pad and bone morphology', *Collegium Anthropol.*, vol. 11, pp. 297–303.

Bidloo, G. (1685) *Anatomia Humani Corporis* (Amsterdam, Utrecht edition 1728), cited by H. Cummins and Ch. Midlo in *Finger Prints, Palms and Soles* (Dover Publ. Inc., 1961).

Bonnevie, K. (1927) 'Die ersten Entwicklungsstadien der Papillarmuster der menschlichen Fingerballen', *Nyt.Mag.f.Naturv.*, vol. 65, pp. 19–56.

Bonnevie, K. (1929) 'Was lehrt die Embryologie der Papillarmuster über ihre Bedeutung als Rassen- und Familiencharakter', *Z.f.indukt.Abst.-u. Vererbungslehre*, vol. 50, pp. 219–74.

Chamla, M. C. (1962–63) 'La répartition géographique des crétes papillaires digitales dans le monde; nouvel essai de synthese', *L'Anthropologie*, vol. 66, pp. 1–47, 67, 526–41.

Chopra, V. P. (1971) 'Biostatistische Analyse der Hautleisten von geistig gestörten Kindern', in: W. Hirsch (ed.) *Hautleisten und Krankheiten*, (Berlin) pp. 341–62.

Faulds, H. (1880) 'On the skin furrows of the hand', *Nature*, vol. 22, p. 605.

Faulds, H. (1905) *Guide to fingerprint identification* (Hanley: Wood, Mitchell and Co.).

Galton, F. (1888) 'Personal identification and description', *J. Anthrop. Inst. London*, vol. 18, pp. 177–91.

Galton, F. (1891) 'Methods of indexing finger-marks', Proceedings of the Royal Society, vol. 49, pp. 540–8.

Galton, F. (1892) *Finger Prints* (London: Macmillan).

Galton, F. (1893) 'Identification', letter in *Nature*, vol. 48, p. 222.

Galton, F. (1895) *Finger Print Directories* (London: Macmillan).

Galton, F. (1896) 'Prints of scars', *Nature*, vol. 53, p. 295.

Geipel, G. (1961) 'Das Tastleistensystem der Hände und die Beugefurchen mongoloider Personen', *Acta Genet.Med.et Gemell* (Roma) vol. 10, pp. 80–92.

Grew, (1684) cited by H. Cummins and Ch. Midlo in *Finger Prints, Palms and Soles* (Dover Publ. Inc., 1961).

Hauser, G. (1986) 'Significance of dermatoglyphics in studies of population genetic variation in man', in D. F. Roberts and G. F. De Stefano (eds), *Genetic Variation and its Maintenance* (Cambridge University Press) pp. 237–43.

Heindl, R. (1927) *System und Praxis der Daktyloskopie und der sonstigen technischen Methoden der Kriminalpolizei* (Berlin and Leipzig: W. De Gruyter).

Herschel, W. (1880) 'Skin furrows of the hand', *Nature*, vol. 23, p. 76.

Jantz, R. L. (1974) 'Multivariate analysis of dermatoglyphic variation in man', *Yearbook of Physical Anthropology*, vol. 18, pp. 121–39.

Jantz, R. L. and D. W. Owsley (1977) 'Factor analysis of finger ridge counts in blacks and whites', *Ann.Hum.Biol.*, vol. 4, pp. 357–66.

Knussmann, R. (1967) 'Interkorrelation im Hautleistensystem des Menschen und ihre faktorenanalytische Auswertung', *Humangenetik*, vol. 4, pp. 221–43.

Knussmann, R. (1973) 'Unterschiede zwischen Mutter–Kind- und Vater–Kind-Korrelationen im Hautleistensystem des Menschen', *Humangenetik*, vol. 19, pp. 145–54.

Kollmann, A. (1883) *Der Tastapparat der Hand der menschlichen Rassen und der Affen in seiner Entwicklung und Gliederung* (Hamburg and Leipzig: L. Voss).

Loesch, D. (1973) 'Minutiae and Clinical Genetics', *J.Ment.Defic.Res.*, vol. 17, pp. 97–105.

Malpighii, M. (1686) *De externo tactus organo* (Londini: Opera omnia II).

Mayer, J. C. A. (1788) *Anatomische Kupfertafeln nebst dazugehörigen Erklärungen* (Berlin: J. Decker and H. A. Rottmann).

Okajima, M. (1967) 'Frequency of epidermal-ridge minutiae in the calcar area of Japanese twins', *Amer.J.of Human Genetics*, vol. 19, pp. 660–73.

Okajima, M. (1975) 'Development of dermal ridges in the fetus', *Journal of Medical Genetics*, vol. 12, pp. 243–50.

Ökrös, S. (1965) *The Heredity of Papillary Patterns* (Budapest: Akadémiai Kiadó – Publishing House of the Hungarian Academy of Sciences).

Penrose, L. S. (1967) 'Finger-print pattern and the sex chromosomes', *Lancet*, vol. i, pp. 298–300.

Plato, Ch. C. (1978) 'The worldwide distribution of dermatoglyphics', in *Human Biology – Recent advances*, vol. 2 (New Delhi: Today & Tomorrow's Printers and Publishers) pp. 365–76.

Poll, H. (1934) *Rasse, Krankheit und Daktylogramm*, Proceedings of Congrès international des Sciences Anthropologique et Ethnologiques, 1st Session, London.

Purkinje, J. E. (1823) *Commentatio de examine physiologico organi visus et systematis cutanei* (Vratislaviae: Typis Universitatis).

Reed, T., J. A. Norton Jr and J. C. Christian (1978) 'Fingerprint Pattern Factors', *Hum.Hered*, vol. 28, pp. 351–60.

Roberts, D. F. and E. Coope (1975) 'Components of variation in a multifactorial character: a dermatoglyphic analysis', *Human Biology*, vol. 47, pp. 169–88.

Roberts, D. F. and E. Coope (1979) 'Digital ridge counts and genetic fields', *Human Genetics*, vol. 47, pp. 151–8.

Schaeuble, J. (1933) 'Die Entstehung der palmaren dipitalen Triradien', *z.Morph.Anthrop.*, vol. 31, pp. 403–38.

Steffens, C. (1965) 'Vergleichende Untersuchungen der Minutiae der Fingerbeerenmuster bei Familien und eineiigen Zwillingspaaren', *Anthrop.Anz*, vol. 29, pp. 234–52.

Thorwald, J. (1965) *Das Jahrhundert der Detektive* (Zürich: Droemer'sche Verlagsanstalt).

Welcker, H. (1898) 'Die Dauerhaftigkeit des Dessins der Riefchen und Fältchen der Hände', *Arch.f.Anthropologie*, vol. 25, pp. 29–32.

11 Galton and Evolutionary Theory

John Maynard Smith

INTRODUCTION

There are two approaches to science: the inductive and the hypothetico-deductive, or, if you prefer, the Baconian and the Popperian. The former measures and records phenomena in the hope that the measurements will lead to a universal descriptive law. The latter starts from a theory or model of the world and measures only to confirm or falsify the model. Real scientists, of course, indulge in both types of activity. The most principled Popperian cannot make a model without some prior knowledge of the world, and Baconians find it hard not to make models to explain their empirical laws. But individual scientists tend to operate in one mode or the other, and sometimes develop explicit philosophies to justify their practice. Karl Pearson and Peter Medawar are examples from different sides of the divide.

Francis Galton was a Baconian. He loved measuring things and was ingenious in inventing new ways of doing so. His main contribution to science lay in the development of statistical methods whereby general descriptive laws may be inferred from measurements. Although he was also ingenious in inventing models, he tended to treat these as analogues rather than as hypotheses: an example is his 'polygon' model of evolution, which is discussed below. I'm afraid this may make me an unsympathetic interpreter of Galton, since I am by habit and prejudice a Popperian. My task is made harder by the fact that Galton's central interest lay in understanding human heredity, whereas I regard the human species as an unsatisfactory experimental animal about which, for largely fortuitous reasons, a lot of information is available. Although Galton had some interesting ideas about evolution – ideas which had an important impact on the later development of evolutionary thinking, particularly in Britain – these arose as a spin-off from his work on heredity, and through family ties with Charles Darwin.

I have therefore chosen to concentrate on three somewhat unconnected topics: his experiments on pangenesis, his ideas about discontinuity in evolution, and, more briefly, his flirtation with Mendelian ratios. I will also comment on the relevance of Galton's work to contemporary evolutionary biology. I know that this lays me open to the accusation of writing Whig history, but that is the only kind of history that a working scientist can be expected to write.

THE EXPERIMENTS ON PANGENESIS

Charles Darwin, like almost all his contemporaries, thought that the characteristics acquired during the lifetime of an individual could influence the nature of their offspring. Partly to account for this supposed effect, he put forward his theory of pangenesis, according to which gemmules (particles carrying what we would today call hereditary information) travel from the various parts of the body to the gonads, where they participate in the formation of the germ cells and hence to the next generation. In 1870–2 Galton, in collaboration with Murie, the prosector at London Zoo, performed a series of experiments to test this idea. They transfused blood from rabbits of various colours – agouti, black-and-white, yellow – into members of a silver-grey breed and then mated the silver-greys among themselves, examining the offspring for signs that the transfusion had affected their colour. In 1871 Galton read a paper to the Royal Society, stating that no such effect had so far been observed, and concluding 'the doctrine of Pangenesis, pure and simple, as I have interpreted it, is incorrect'. He continued experiments, with improved techniques, into 1872, but the results continued to be negative.

Darwin, not surprisingly, was disappointed by these negative results but, also not surprisingly, he clung to his theory of pangenesis. He wrote to *Nature*, claiming that 'I have not said one word about the blood, or about any fluid proper to any circulating system. It is, indeed, obvious that the presence of gemmules in the blood can form no necessary part of my hypothesis, for I refer in illustration of it to the lowest animals, such as the Protozoa, which do not possess blood or any vessels; and I refer to plants in which the fluid, when present in the vessels, cannot be considered as true blood'. However Galton had every reason for supposing that Darwin's theory included transport of the gemmules in the blood. Darwin had written 'the gemules in each organism must be thoroughly diffused; nor does this seem improbable, considering their minuteness, and the steady circulation of fluids throughout the body'. More importantly Galton had kept Darwin fully informed of his experiments, and solicited his advice. If Darwin had really not expected the gemmules to be transported in the blood, he could easily have said so. Galton subsequently behaved with considerable magnanimity. In a letter to *Nature*, replying to Darwin, he accepts that he may have misunderstood Darwin's meaning.

It is instructive to compare Galton's experiment with Weismann's famous one in which he cut the tails off mice and found that their offspring were born with perfect tails. The design of Galton's experiment suggests that its author hoped, and perhaps expected, that it would give positive results. Galton's correspondence with Darwin makes it clear that this is the case. For example, on 15 March 1870 he wrote 'For my part I am quite sick with expected hope and doubt'. On 19 March Emma Darwin wrote to her daughter 'F. Galton's experiments about rabbits [viz. injecting black rabbit's blood into grey and vice versa] are failing, which is a dreadful disappointment to both of them'. On 31 March, when some signs of positive results were apparent, Galton's letter to Darwin opens 'Better news – decidedly better'.

In contrast Weismann's experiment is clearly intended to fail. I first learnt of it when, as a schoolboy, I read the preface to *Back to Methuselah* (1921), in which Shaw attacks Weismann for cruelty and stupidity. No Lamarckist, Shaw argues, would expect surgical injuries to be inherited: only actively acquired adaptations should be transmitted. Much later, on reading *The Evolution Theory* (Weismann, 1904), I discovered that Weismann had better reasons for rejecting the inheritance of acquired characters than the negative results of his experiments, and better reasons for doing the experiment than Shaw allowed. His rejection was based on a number of theoretical arguments, of which two are particularly persuasive. First, many complex adaptations could not have evolved by a Lamarckian mechanism because they could never have arisen as individually acquired adaptations in an ancestor. Among the examples he quotes are the adaptations of sterile castes of social insects; of the exoskeleton of insects, which is hardened and fixed in form before it is used; and of the egg-laying behaviour of butterflies and moths, which die before the success or failure of their oviposition site has been decided. These examples show that Weismann, like Darwin and Wallace but unlike Galton, thought as a naturalist. Of course the argument shows only that complex adaptations can evolve without

Lamarckism – it does not prove that acquired characters cannot be transmitted.

Weismann's second and more fundamental argument was curiously modern. He argues that if, for example, a learnt behaviour is to become instinctive, then information about the change in the structure of the brain that has occurred during the life of an individual must be translated into a different form so that it can be transmitted in the gametes: he was unable to conceive how this could happen. It is interesting that the current debate about the latest 'Lamarckian' claim turns on exactly this point. Experiments by Cairns et al. (1988) and others suggest that, in some circumstances, mutations in bacteria are not random, but specific to the needs of the cell. The matter is still controversial, but everyone agrees that final acceptance of the claim will depend on the discovery of a mechanism whereby the mutational process can be altered in an adaptive way by the state of the cell: several possible mechanisms have been suggested.

Weismann, then, differed from Galton in expecting his experiment to fail. Why then did he do it? He describes how, when he first proposed that acquired characters were not inherited, he was met with the objection that it was known that mutilations were inherited – for example the birth of tailless puppies to bitches whose tails had been docked. He adds 'even students' fencing scars were said to have been occasionally transmitted to their sons (happily, not to their daughters)'.

In assessing this story it is important to recognise, as Weismann himself admitted, that Galton was a 'Weismannist' before Weismann, and that he was sceptical of the inheritance of acquired characters before his pangenesis experiments. His hopes for a positive result in these experiments may have owed more to his friendship with Darwin than to a belief in the phenomenon being tested. Nevertheless his experiment does look to me as if it was intended to succeed. Galton, I think, designed the better experiment, but Weismann had a deeper understanding of the problem.

REGRESSION, VARIATION AND SPORTS

Galton's studies on heredity led him to draw a sharp distinction between two types of variation: in the 1892 reprint of *Hereditary Genius*, he wrote 'the word variation is used indiscriminately to express two fundamentally distinct conceptions: sports and variations properly so called'. This was a distinction he returned to repeatedly. By 'variations properly so called' he meant the type of variation he had himself studied in seed size in sweet peas and in human stature – that is, what we would today call 'continuous variation'; by a 'sport' he meant the sudden appearance of a qualitatively new type, or of a type outside the range of existing continuous variation. He held that evolutionary change depended on the natural selection of sports, and not of variations. He illustrated this conclusion by his analogy of a polygon that can stand on any of its edges on a flat surface: 'The model and the organic structure have the cardinal fact in common, that if either is disturbed without transgressing the range of its stability, it will tend to re-establish itself, but if the range is over-passed it will topple into a new position. . . . The ultimate point to be represented is this. Though a long-established race habitually breeds true to its kind, subject to small unstable deviations, yet every now and then the offspring of these deviations do not tend to revert but possess stability of their own'.

What led him to this curious position, and what were the consequences of his holding it? He had the idea of a reversion or recoil towards a point of stability before he had any strong evidence to support it. Thus although the quotation in the last paragraph is from *Natural Inheritance*, published in 1889, he used a very similar analogy in 1869 in *Hereditary Genius*. It would be interesting to trace the source of this idea: it may have owed something to his own observation that great gifts are not transmitted indefinitely. In any case the importance of this and other theoretical ideas for his later work on heredity suggest that Galton was not as consistent an inductionist as I suggested at the start of this essay. His belief in the idea of reversion was greatly strengthened by his study of seed size in sweet peas, reported in 1877. He observed that if he plotted the seed size of offspring against the average seed size of the two parents (that is, the mid-parent), the slope of the graph was 2/3, and not 1: that is, the offspring mean was closer to the mean of the population than was the mean of the parents. This was the origin of the concept of 'regression'. More precisely it was the origin of two quite different concepts: it was the first use of the method of plotting regression lines, which was an important landmark in the history of statistics, and it provided empirical support for the idea that regression towards the mean was an important phenomenon in heredity, which seems to me to have had wholly malign effects.

If, today, one observes a 'realised heritability' of less than unity –

as of course one almost invariably does – there are two possible explanations one can offer: either there are non-additive gene effects (epistasis, dominance), or there are environmental causes affecting the measured trait that do not act identically on parents and offspring. Usually analysis shows that both types of effect are present. It is hardly surprising that Galton did not consider non-additive gene effects, although both he and Darwin had the idea that a gemmule could be 'latent' in an individual but be expressed in a descendant – an idea analogous to Mendel's concept of recessivity. I find it odd, however, that neither Galton nor Pearson, in his long and arcane discussion of Galton's findings, seem to have considered the environmental explanation. Although both were fairly extreme hereditarians, both must have been aware that some differences are environmentally caused. I take it that the difficulty lay in the absence of an adequate theory connecting causation, correlation and (statistical) regression. Today a realised heritability (that is, the slope of offspring mean on mid-parent) is regarded as a measure of the 'additive genetic variance' (that variance caused by the additive effects of genes) divided by the total variance (the sum of the genetically and environmentally caused variances). In the absence of such a way of looking at things, a regression slope of 2/3 must have been difficult to interpret.

In fact Galton interpreted it in terms of a 'law of ancestral inheritance'. The hereditary characteristics of the child were held to depend not only on its parents, but on its grandparents, and to a decreasing degree on all its ancestors, however distant. Galton thought that this more distant ancestral influence would cause the regression towards the mean that he observed: 'The child inherits partly from his parents, partly from his ancestry. Speaking generally, the further his genealogy goes back, the more numerous and varied will his ancestry become, until they cease to differ from any equally numerous sample taken at haphazard from the race at large. Their mean stature will then be the same as that of the race; in other words, it will be mediocre'.

As a consequence of this way of thinking, the population mean took on for Galton the characteristics of an unchanging centre to which the population must always return. Natural selection, acting on mere variations, cannot permanently alter a population: 'mere varieties from a common typical centre blend freely in the offspring, and the offspring of every race whose *statistical* characters are constant, necessarily tend, as I have often shown, to revert to their common typical centre'. Since Galton was an evolutionist he was forced to suppose that the genetic changes relevant in evolution – sports – were different in kind from those that underlay the variability of populations. It is hard to imagine a more unfortunate conclusion.

Although this opinion of Galton's was reached from a study of continuous variation, it was confirmed by his work on fingerprints. He concluded that there were qualitative differences between 'arches', 'loops' and 'whorls', different in kind from the continuous variation within each class: he held this view despite the occurrence of intermediates between these classes that were hard to classify. He speaks of these differences as being similar to those between 'genera'. Since the same individual can have an arch on one thumb and a loop on the other, I take it that he meant no more than that the differences are qualitative. Unfortunately some of the later Mendelians – De Vries for example – came to see the origin of a new species as equivalent to a sport, or mutation.

Galton's conviction that variations and sports were fundamentally different had unhappy consequences for evolutionary biology, particularly in Britain. Two men, Karl Pearson and William Bateson, both strongly influenced by Galton, became the leaders of two schools, the biometricians and the Mendelians, who successfully muddied the waters of evolutionary thought until some clarity was brought to the field by R. A. Fisher and J. B. S. Haldane. Pearson thought that Galton had erred in his interpretation of regression. He argued that there was no necessity for a population to regress towards the mean. Natural selection, acting on variations, could alter a population and could account for evolution. Hence there was no need to ascribe a special evolutionary role to sports. Philosophically, Pearson held that the business of science is to describe the world but not to postulate the existence of entities – whether atoms or genes – that cannot directly be observed. Genes, therefore, and gene mutations, are not only unnecessary hypotheses, but contrary to the proper method of science. Nevertheless Pearson saw himself as the heir of Galton because he shared his faith in eugenics, and because he developed Galton's statistical methods.

Although Pearson's claim to be Galton's heir was justified, since he built on Galton's positive achievements, Bateson was closer to Galton's views as far as evolution was concerned. In his *Materials for the Study of Variation* (1894), Bateson argued convincingly that much morphological variation was discontinuous. Differences similar to those between arches, loops and whorls occur repeatedly, both within and between species. He was therefore ready to appreciate Mendel's laws when they were rediscovered, and he became a leading figure in the development of Mendelian genetics. Whereas Pearson denied the evolutionary relevance of sports and regarded continuous variation as the raw material of evolution, Bateson, like Galton before him, regarded continuous variation as unimportant in evolution and sports as essential: he went further, in denying that natural selection played any significant role. The tragedy of course is that both men continued to regard the two types of variation as fundamentally distinct: this was a legacy from Galton.

Attitudes to continuous and discontinuous variation in the nineteenth century deserve further study. David Burbridge has pointed out to me that much confusion arose because of a failure to distinguish clearly between (a) the inherent restriction of a variable to discrete values, and (b) a gap in the observed statistical distribution of a trait, such as height, which is inherently capable of varying continuously. He suggests that Pearson may have been the only writer in the last century to be reasonably clear on this issue. Galton, when speaking of sports and variations, seems usually to have had a distinction of type (b) in mind, but he is inconsistent. For example he quotes topological variation in fingerprint patterns in support of the distinction.

Then as now, Darwin's insistence on gradual change was crucial. Gruber (1974) has argued that Darwin acquired the idea 'gradual = natural: sudden = miraculous' from the theologian Sumner when he was an undergraduate and a believing Christian, but that the idea survived when he became an agnostic and an evolutionist in the form of a conviction that a natural explanation of evolution requires that complex adaptations arise in a number of steps, each naturally selected, and not in a single large step, which would need miraculous intervention. This conviction would be shared by most evolutionists today – it does not rule out either type of discontinuous change.

The difficulty of reconciling continuous and discrete models of the world is still with us. With wave-particle duality, physicists have accepted the need, at the most fundamental level, for the coexistence of both types of model. For geneticists the fundamental level is discrete. As Haldane wrote in *The Causes of Evolution* (1932), 'Here matters would have been easier if heritable variations really formed a continuum, as Darwin apparently thought, *i.e.* if there was no limit to the possible smallness of a variation. But this is clearly not the case when we are considering meristic variation. Mammals have a definite number of neck vertebrae and chromosomes, most flowers a definite number of petals, exceptional organisms being unhealthy. And the atomic nature of Mendelian inheritance suggests very strongly that even where variation is apparently continuous this appearance is deceptive. On any chemical theory of the nature of genes this must be so'. However, since genes code for proteins and some proteins influence the phenotype by altering the rates of the chemical reactions they catalyse, and since a single aspect of the phenotype is usually affected by many genes, it is not surprising that much phenotypic variation appears continuous.

The recognition that discrete genetic differences underlie apparently continuous phenotypic variation has been with us for over fifty years. More recently we have become aware that effectively continuous variation in underlying parameters may be responsible for the meristic variation referred to in the quotation from Haldane. We now know that, as one gradually alters the parameters of a dynamical system, its behaviour may alter suddenly and qualitatively. There is a mathematical language in which we can talk about such phenomena. Although few of us properly understand the underlying mathematics, we are increasingly familiar with the phenomena, in part from computer simulation. Our ignorance of developmental biology is still such that we cannot give a detailed account of the meristic and topological variation that fascinated Bateson and, in the case of fingerprints, Galton, but we no longer think that the genetic causes of meristic and continuous variation are fundamentally distinct.

Marx and Engels, who wanted to talk about historical change, borrowed from Hegel the concept of the transformation of quantity into quality. In the absence of a more precise mathematical language they were wise to do so. Unfortunately no comparable borrowing took place in evolutionary biology. From Darwin's gradualism to the saltationism of the Mendelians, continuous and discontinuous change seemed fundamentally distinct. Galton only made explicit a difficulty that everyone seems to have felt.

Real disagreements remain, however, as will be apparent from the recent debate about 'punctuated equilibria'. There is still room for doubt about the importance of discontinuous phenotypic variation of the kind discussed by Bateson, even if we no longer think that it has a different genetic basis from continuous variation. Perhaps more fundamental is the debate about how genotypic change maps onto fitness. At one extreme is the Fisherian view that evolution consists of hill-climbing on a smooth fitness surface. If so, natural selection is the main driving force, and large random-mating populations, with their copious supply of mutations, can evolve most rapidly, although of course they will not do so if they are already close to a fitness optimum. Rapid evolution will result from a change in the environment, and hence in the fitness surface. At the other extreme is the Wrightian view, according to which evolution requires the crossing of valleys in the fitness surface. A large population subdivided into partially isolated demes is the most effective evolutionary unit because of the opportunities it provides for the chance crossing of valleys.

A PRIORI MENDELISM

There is one curiosity in Galton's career that I cannot refrain from commenting on. In his correspondence with Darwin on pangenesis, he almost stumbled on Mendel's laws *a priori*. Darwin had written to him and asked how one could explain the fact that, in many hybrids, every part is intermediate between the parents, and 'this hybrid will produce by buds millions on millions of other buds all exactly reproducing the intermediate character'. In his reply Galton argues that if the gemmules themselves retained their unique characteristics (for example black and white), the cells of the hybrid would be intermediate (that is grey) if they contained equal numbers of the two types. He continues: 'The larger the number of gemmules in each organic molecule, the more *uniform* will the tint of greyish be in the different units of structure. It has been an old idea of mine, not yet discarded and not yet worked out, that the number of units in each molecule may admit of being discovered by noting the relative number of cases of each grade of deviation from the mean greyness. If there were 2 gemmules only, each of which might be either white or black, then in a large number of cases one-quarter would always be quite white, one-quarter quite black, and one-half would be grey. If there were three molecules, we should have 4 grades of colour (1 quite white, 3 light grey, 3 dark grey, 1 quite black and so on according to the successive lines of "Pascal's triangle"). This way of looking at the matter would perhaps show (a) whether the number in each given species of molecule was constant, and (b), if so, what those numbers were'.

He seems to be imagining that, in the case of two gemmules per 'molecule', each molecule is formed by the random fusion of two gemmules. What a tragedy that he did not apply his idea, not to the cells, or 'molecules', of the F1, but to the individual organisms in the F2!

As in the pangenesis experiments, there is an intriguing parallel between Galton and Weismann. The latter also came close to discovering Mendelism *a priori*. Figure 87 of *The Evolution Theory* shows the expected constitution of the zygote in successive generations of outcrossing. Unluckily he assumes that there are eight ids (genes) in a gamete and sixteen in a zygote: it must have seemed too simple to assume diploid inheritance. There are, I think, other confusions in his diagram. It is an interesting comment on the methods of biology in 1900 that this is the only diagram in the book although there are over 100 figures and although Weismann was an inveterate theoriser. Model-building was not part of the practice of professional biologists at that time. Despite his inductionist tendencies, Galton, the amateur, was in some ways closer to modern methods of working.

GALTON AND EVOLUTION

Galton's major research interests were in human heredity and in the measurement of human variation: his religion, insofar as he had one, was eugenics. His interest in evolutionary biology was secondary, deriving from his interest in heredity. He lacked the passion for natural history that was a driving force for Darwin, Wallace and Weismann. He did, however, share the other major motive that has aroused people's interest in evolution – the hope that an understanding of evolution can liberate us from the need to believe in a creator. In a letter to Darwin in response to the latter's congratulations on *Hereditary Genius*, Galton wrote 'I always think of you in the same way as converts from barbarism think of the teacher who first relieved them of the intolerable burden of superstition. I used to be wretched under the weight of the old-fashioned arguments from design, of which I felt, though I was unable to prove to myself, the worthlessness. Consequently the appearance of your *Origin of Species* formed a real crisis in my life; your book drove away the constraint of my old superstition.'

Acknowledgement

An earlier version of this essay was read by Mr David Burbridge, who made a number of suggestions, most of which I have accepted.

References

Bateson, William (1894) *Materials for the Study of Variation, Treated with Especial Regard to Discontinuity in the Origin of Species* (London: Macmillan).

Cairns, J., J. Overbaugh and S. Miller (1988) 'The origin of mutants', *Nature*, vol. 335, pp. 142–5.

Galton, F. (1869) *Hereditary Genius, an Inquiry into its Laws and Consequences* (London: Macmillan).

Galton, F. *Natural Inheritance* (1889) (London: Macmillan).

Gruber, H. E. (1974) *Darwin on Man* (London: Wildwood House).

Haldane, J. B. S. (1932) *The Causes of Evolution* (London).

Pearson, K. (1924, 1930) *The Life, Letters and Labours of Francis Galton*, Vols 2 and 3 (Cambridge University Press).

Shaw, George Bernard (1921) *Back to Methuselah* (New York).

Weismann, A. (1904) *The Evolution Theory* (trans. J. A. and M. R. Thomson) (London: Edward Arnold).

12 Galton's Conception of Race in Historical Perspective

Michael Banton

Like many of his contemporaries, Galton used the word race in a variety of senses. First of all there was a literary sense, as when he wrote that 'judges are by no means an unfertile race', that 'poets are a sensuous, erotic race, exceedingly irregular in their way of life', and that there was 'no reason to believe that Divines are an exceptionally favoured race' (1869, pp. 131, 225, 274). Secondly there are passages in which he appears to use the word in a taxonomic sense. It looks as if he is using race as a synonym for genus when he speaks of 'the human race'; as a synonym for species when he writes that 'the American Indian race is divided into many varieties'; and possibly as a synonym for variety when he refers to the race of the Irish Celt, or to 'our race' when he seems to mean the English. However he draws no distinction between a race and a breed and can refer to 'a highly bred human race' (1865, pp. 319–20, 1869, p. xxiv). Thirdly, and in his later writing rather than in that of the 1860s, there is a sense in which race is used apparently as a synonym for heredity. What otherwise is one to make of the passages in the 1892 preface to *Hereditary Genius* in which he writes about 'the question of race', 'the influence of race' and the 'importance to be attached to race', going on to regret scientific ignorance of the 'respective ranges of the natural and acquired faculties in different races'? Since this is an elaboration of 'the importance to be attached to race', it seems as if the third and second senses feature in successive sentences.

TAXONOMY

Many anthropologists of Galton's generation were very much concerned with the taxonomic problem, and disputed about where race fitted into the sequence of genus, species and variety (indeed the furore over what has been called the scientific racism of the 1850s and

1860s was very largely a dispute over whether race was to be equated with species or variety). There is no indication that Galton took much interest in this, which is why I said only that he appears to use the word race in a taxonomic sense. Galton's focus was on race as a line of inheritance. If we set aside the literary sense, as in his reference to judges, this means that his apparent references to race as a taxon, the second sense, are better read in the third sense as referring to taxa constituted by inheritance. Galton's neglect of taxonomy is the more unfortunate since it is the source of so much confusion in this field. One of the earliest senses of the word in English was to denote a taxon constituted by descent, as in a 1570 reference to 'the race and stocke of Abraham', but in the mid-eighteenth century Linnaeus constituted taxa by characters at one point in time and shortly afterwards Latin nomenclature began to give way to vernacular names. Thus 'race' was used to denote a Linnaean taxon. In 1784–5 both Kant and Herder protested about this extension in the use of the word race in German. They protested in vain, because in 1817 Cuvier lent his great authority to the equation of race with variety as a taxon. The characteristics shared by the individual specimens assigned to a variety were presumed to derive from their descent. Instead of looking to see whether individuals known to share ancestry had common characters, the argument was reversed. Ancestry was assumed to explain shared characters. Cuvier also introduced the concept of type as a taxon of uncertain status and stressed its stability. Thus there grew up a pre-Darwinian theory of human variation best formulated by Louis Agassiz. He divided the world into eight natural provinces for each of which there was a distinctive and unchanging human species, as could be deduced from their present-day characters. Whereas the Linnaean classification was based on physical characters only, the typological school with which Agassiz identified himself constituted human taxa on the basis of cultural or behavioural as well as physical characters.

With the advantages of hindsight we can say that the era of Cuvier ended in 1859. In *On the Origin of Species* (1859), Darwin referred to 'geographical races or subspecies' as 'local forms completely fixed and isolated'. Because they were isolated they did not interbreed, and so 'there is no possible test but individual opinion to determine which of them shall be considered as species and which as varieties'. As Darwin was attempting to explain the way living things changed, questions of classification at moments of time were not of central importance. Race was a taxon constituted by descent but its individual members were not identical. The theory of natural selection as advanced by Darwin left many important questions unresolved. It demanded a greater conceptual reorientation than was apparent at the time and was more complex than it seemed. So many people, scientists as well as non-scientists, continued for a long time to think about race in terms that we can now see to be pre-Darwinian. Concepts also are subject to a process of selection, so that as far as scientific explanation is concerned we should expect those senses of the word race which are not useful to be gradually discarded. If they have a political use, however, they may start on a new career.

Galton understood Darwin's teaching better than most people. Ernst Mayr (1982, p. 697) describes him as 'a strong proponent of population thinking, appreciating the uniqueness of the individual more clearly than any of his contemporaries. This led him to his discovery of the uniqueness of fingerprints . . .' As his interest in the comparison of individuals developed he came to say that race 'meant no more than the average of the characteristics of all the persons who were supposed to belong to the race, and this average was continually varying. The popular notion seemed based upon some idea like that of a common descent of the different races, from a parent Noahican stock, whence the aborigines of each country were derived . . .' (Galton, 1881, pp. 352–3). Popular suppositions about belonging to one race rather than another seem not to have been important. They only described the sample. This helps explain his neglect of the taxonomic problem.

Some would interpret Galton's references to race in a more hostile way, as building blocks in a structure of scientific racism which, unlike Agassiz' version, rationalised a hierarchy of superiority and inferiority. In *Hereditary Genius*, by incorporating gross and unreliable estimates in his calculation, he presented Athenian Greeks as on average two grades (out of eight) above 'our own', 'the negro' as two grades below and the 'Australian type' at least one further grade lower. In 1872 he wrote to *The Times* to argue for the introduction of Chinese into Africa. Though a temporary dark age prevailed in China it had not sapped the genius of the race. 'The Chinese emigrants possess an extraordinary instinct for political and social organisation, they contrive to establish for themselves a police and internal government, and they give no trouble to their rulers so long as they are left to manage these things for themselves'. In Africa, one population continually drives out another. A new competitor, the Chinaman, could be introduced. 'The gain would be immense to the whole civilized world if he were to outbreed and finally displace the negro, as completely as the latter has displaced the aborigines of the West Indies' (quoted Pearson, 1924, vol II, p. 33). Six years later he wrote of 'the negro' that 'His coarse pleasures, vigorous physique, and indolent moods, as compared with those of Europeans, bear some analogy to the corresponding qualities in the African buffalo, long since acclimatized in Italy, as compared with those of the cattle of Europe. Most of us have observed in the Campagna of Rome the ways of that ferocious, powerful and yet indolent brute' (ibid, pp. 31–2). This was the vein in which Victorians often generalised, but it may not have been a good predictor of how they behaved in actual situations. Thus at the Royal Geographical Society in 1858, commenting on a later traveller's account of a people Galton had visited, he criticised his successor's conduct in strong terms as aggressive, remarking 'I was able to leave, in peace, the happy country of a noble and kindly negro race, which has now, for the first time, been confronted and humbled before the arrogant strength of the white man' (reprinted in Galton, 1889, p. 199). Later he declared that it was 'a shame to us as an Imperial nation, that representatives of the many people whom we governed, did not find themselves more at home among us . . . this [was a] serious drawback to our national character as rulers of a great Empire'. He told the Anthropological Institute that 'Anthropology teaches us to sympathize with other races, and to regard them as kinsmen rather than as aliens. In this respect it may be looked upon as a pursuit of no small political value' (Galton, 1881, p. 353; 1885, pp. 337–8).

EVOLUTION

Karl Pearson considered that the foundation stone of Galton's anthropological work was his belief that both physical and psychical characters were inherited; these two together constituted nature or 'race' as opposed to nurture. In 1873 Galton was contending that 'Race has a double effect, it creates better and more intelligent individuals, and these become more competent than their predecessors to make laws and customs, whose effects shall favourably react on their own health and on the nurture of their children' (Pearson, 1925, vol. II, pp. 75, 117–18). His example of favourable nurture is sanitary improvements. He says remarkably little about the laws and customs. It is from this standpoint that we should consider his Utopia, sketched already in that passage in *Hereditary Genius* in which he described 'the best form of civilization in respect to the improvement of the race'. It was one 'where the pride of race was encouraged (of course I do not refer to the nonsensical sentiment of the present day, that goes under that name)'. To what, then, did he refer? A concern for eugenic mating, sanitary improvement, and, surely, a sense of team-spirit?

Darwin had recognised the importance of cooperation and altruism when he wrote that 'I use the term Struggle for Existence in a large and metaphorical sense, including dependence of one being on another' (1859, p. 62). He and Galton both appreciated the importance of competition between societies and Galton contended that a shared religion, whatever its character, could be very important to the functioning of a society (Pearson, 1930, A: 88). The sociobiologists have since assembled evidence indicating that some of this ability to cooperate is inherited genetically, but much is acquired culturally and is a product of particular social arrangements in which laws and customs are crucial. So in Utopia it was necessary that the pride of race should be encouraged by social action. There is here a deficiency in Galton's vision of society which is the more surprising because other members of the learned societies to which he belonged were constantly discussing variations in such matters as the prohibition of incest, rules of exogamy and the reckoning of descent. Sir Henry Maine, Galton's friend at Cambridge, was contending that social progress depended on getting the right relation between law and custom, and in particular on a movement from relations based on status to relations based on contract, to an order in which relations arose from the free agreement of individuals. The opportunity to enter into contracts was vital to the development of cooperation, while the rule of law had made possible that sudden spurt in economic and social development from which those of Galton's generation and class had benefited so much. Galton's own theories could not account for it.

There was another way of looking at human variation which did seem to account for that advance. This was the thesis that race in the sense of inheritance was a prime determinant of culture or level of civilisation. It was the more attractive because of the ambiguity in the meaning of the word race. Stocking (1987, pp. 138–9, 143, 236–7) notes that for Leslie Stephen in 1900 the word could serve as 'a kind of summation of historically accumulated moral differences sustained and slowly modified from generation to generation'. Cultural differences were thought to be the products of environmental influences, but they were also seen as somehow inherited and therefore en-meshed in a framework of biological evolutionism for which race had become a shorthand designation. More and more of what might be accounted nurture was being packed in with nature so that it was starting to include forms of cooperation. Even specialists continued to use race in the sense of inheritance alongside race as a taxon. Thus G. Elliot-Smith (1924; 1968, p. 26) in his Galton lecture to the Eugenics Society in 1924, after questioning whether 'the form assumed by culture is wholly or primarily a question of race', im-mediately added 'the varying temperaments of races are patent enough'. Elliot-Smith was addressing what in cultural anthropology has come to be called 'Galton's Problem'. Sir Edward Tylor had addressed the Anthropological Institute on the development of social institutions associated with marriage and descent. It suggested an evolutionary trend. Galton, in the chair at the meeting, objected that the institutions could have spread by diffusion and that Tylor's cor-relations might be explained by this intrusive variable. In recent times seven solutions to the problem have been proposed (Naroll, 1970; Driver and Chaney, 1970).

The arrangement of plants and animals in taxa was an objective process of classification. Those who attempted the same for humans could claim that their methods were no different; but they overlooked something. The classification of specimens may have implications for the way they are treated and when those specimens are human, and know how they are being classified, they may wish to dispute the implications. The nineteenth-century classification of humans was bound to be interpreted within an evolutionary framework. An En-glishman could conclude that he belonged in the most advanced race, from which derived special rights and duties. It was for him to command others for their benefit. It was his duty to take up the white man's burden. From what were supposed to be judgements of fact, highly political judgements of value were derived.

My thesis is that the objective and subjective dimensions of racial classification were severed and reformulated with the establishment of population genetics. In 1930 Sir Ronald Fisher's book on *The Genetical Theory of Natural Selection* identified the gene rather than the species as the unit of selection and attributed to each gene a particular fitness value. It led to a view of evolution as the change of gene frequencies in populations. A population characterised by a particular gene frequency could still be called a race, but so designat-ing it did not add any meaning. For biological purposes the concept of race was superseded by a battery of more powerful concepts. The 1935 book *We Europeans* by Sir Julian Huxley and A. C. Haddon can be seen as starting a comparable revision of concepts for handling the subjective dimension of racial classification. The authors emphasised that in the formation of races or subspecies among animals, stability of habitat was a basic factor, whereas humans were continually migrating and mating with new kinds of partner. Since the human population of any region would include many combinations of genes the physical characteristics could be described only statistically. As tribes, nations or peoples, these populations in their political aspect were better designated ethnic groups. Tribes, nations and peoples exist because their members choose to identify themselves with their fellows as constituting such groups. Their classification and nomenclature should reflect this.

That commentators could be slow to appreciate the significance of population genetics for the conceptualisation of race is illustrated by a pamphlet on *West Indian Immigration* published in 1958 by the Eugenics Society after discussion in that society's council. It was written by its general secretary, who argued for research into what he called 'race mixing' or 'miscegenation', and for a recognition that this more commonly led to trouble than to happiness. In its implication that there can be unmixed races, this was pre-Darwinian talk. Also it is at least ironic that the very word miscegenation was coined during the 1864 US presidential election campaign by the dirty tricks department of the Democratic Party in an attempt to identify their opponents with advocacy of such a practice. The 1958 pamphlet referred to 'the distaste which affects many of unpigmented skin at the idea of a white girl breeding with a person whose epidermis contains black granules' as an emotion of group-protection adjusted to a species in evolution, and it was fearful of an emotional and genetic chaos (Bertram, 1958, p. 18). By this time there was plenty of evidence, some of it in books listed in the pamphlet's bibliography, that white sentiments about black–white mating were the result not of genetic inheritance but of social conditioning. Gallup Poll surveys in Britain have reported that the acceptability to whites of marriage between blacks and whites more than doubled between 1964 and 1981.

Galton's Utopia was one in which 'the better sort of emigrants and refugees from other lands were invited and welcomed, and their descendants naturalized'. Immigrants in Britain and other European countries during the last forty years have done much for the receiving countries. Without them the birthrate in many countries would have fallen below the net reproduction rate. By their readiness to move to the places where there is work, to change jobs, work unsocial hours and perform the unattractive functions, often in the face to hostility, they have shown how difficult it is to decide who count as the better sort of people.

CONCLUSION

The conception of race as an objectively constituted taxon has been replaced in the social realm by a new set of concepts that help explain the behavioural processes which constitute racial and ethnic categories and groups. They illustrate the importance of the subjective dimension to the biosocial fabric. Many black, brown, white and yellow people draw upon nineteenth-century racial ideas to classify others whom they see around. Racial assignment can contribute to conceptions of self and can influence behaviour towards others. Self-conception is illustrated by the classification approved by the government for ethnic monitoring schemes and used in the 1991 census, because it invites individuals to assign themselves to one of nine categories. Twenty years ago these would probably have been called racial, but since the process is based on self-assignment it is more appropriately called ethnic in line with the Huxley and Haddon view of ethnic groups as self-constituting political units. Subjective classification is also a defining feature of direct discrimination in law. If one person treats an applicant for a job or for housing less favourably on the ground of the other's presumed race, the action is unlawful even if the first person was quite wrong about the other's presumed race. The persistence of ideas about racial categories which define social rights and duties is part of the process whereby boundaries of exclusion are created and maintained. The then home secretary decided that the British anti-discrimination agency should be called the Commission for Racial Equality, despite acknowledged objections to the adjective, because he considered that government action must be designed to counter mistaken beliefs among ordinary men and women for whom a belief about racial differences had become a ground for invidious distinction.

We can now understand, in a way that no-one of Galton's generation ever could, that the biosocial fabric is self-constituting. There are physical differences of age, sex, pigmentation and relations of kinship which people vest with a social significance that varies in time and space. Many of our contemporaries, when asked, say they are not satisfied with the shape of their bodies; they vary their diet or daily regime to try to get closer to a cultural ideal. Their behaviour as consumers and electors is influenced by similar considerations. The use of their resources to achieve such goals is part of what Galton would have accounted nurture; it influences their nature in more ways than he allowed for. In our studies of the biosocial fabric we now have better tools for examining nature, nurture and their interaction. We should avoid older expressions which may encourage popular misconceptions.

It has been asked whether Galton was a racist? This question does not deserve any lengthy answer, since it is meaningful only if it was possible for Galton to be either a racist or not a racist. If by 'racist' is meant someone who subscribed to a particular explanation of human variation, then this has to be answered against the background of the scientific knowledge of the time. It is a question asked of a group rather than of an individual. Galton knew, better than most, that there was no permanence of type, but his vision of social processes was blinkered and some of his writing displayed a bias of his nation, his class and his generation. Usually, however, the description 'racist' implies moral disapproval. The accuser claims to be morally superior to the accused in the relevant respect. If our predecessors are to be criticised in this way, we should remember that we in turn may be criticised by our successors for moral failings of which we are quite unaware, and which we therefore do not try to correct.

References

Banton, M. (1987) *Racial Theories* (Cambridge University Press).
Bertram, G. C. L. (1958) *West Indian Immigration* (London: The Eugenics Society).
Darwin, C. (1859) *On the Origin of Species* (London: Murray).
Driver, H. E. and R. P. Chaney (1970) 'Cross-Cultural Sampling and Galton's Problem', in N. Naroll and R. Cohen (eds), *A Handbook of Method in Cultural Anthropology*, pp. 990–1003, (New York: Columbia University Press).
Elliot-Smith, G. (1924) 'Problems of Race', reprinted in 1968 in *Eugenics Review*, vol. 60, pp. 25–31.
Fisher, R. A. (1930) *The Genetical Theory of Natural Selection* (Oxford: Clarendon).
Galton, F. (1865) 'Hereditary Talent and Character', *Macmillan's Magazine*,

vol. 12, pp. 157–66, 318–27. Second Part reprinted 1979 in M. D. Biddiss (ed.), *Images of Race*, pp. 57–71 (Leicester University Press).

Galton, F. (1869) *Hereditary Genius*, reprinted 1978 with preface from 1892 edition (London: Julian Friedman).

Galton, F. (1881) 'Discussion of Sir H. Bartle Frere's paper on the Relations between Civilized and Savage Life', *J. Anthropological Inst.*, vol. 11, pp. 352–4.

Galton, F. (1885) 'Remarks made at the commencement of the session', *J. Anthropological Inst.*, vol. 15, pp. 336–8.

Galton, F. (1889) *Narrative of an Explorer in Tropical South Africa* (London: Ward, Lock).

Huxley, J. S. and A. C. Haddon (1935) *We Europeans: A Survey of Racial Problems* (London: Cape).

Mayr, E. (1982) *The Growth of Biological Thought* (Cambridge Mass: Harvard University Press).

Naroll, R. (1970) 'Galton's problem', in R. Naroll and R. Cohen (eds), *A Handbook of Method in Cultural Anthropology*, pp. 974–89 (New York: Columbia University Press).

Pearson, K. (1924, 1930) *The Life, Letters and Labours of Francis Galton*, vols II and IIIA (Cambridge University Press).

Stocking, G. W. Jr. (1987) *Victorian Anthropology* (London: Collier-Macmillan).

13 Galton, the Educationist

W. H. G. Armytage

Galton's friendship and business association with Thomas Hughes (1822–96), the author of *Tom Brown's Schooldays* (1857) – surely the most famous school story written in England – was literally literary: they were both members of the editorial board of a weekly periodical called *The Reader*, which aimed to supply the need for 'a weekly organ which would afford scientific men a means of communication between themselves and the public'.

The first meeting of the original shareholders was held in Thomas Hughes's rooms in Lincoln's Inn Fields on 15 November 1864. Two years later *The Reader* collapsed owing to the multiple clashing of egos, and three years later another periodical with the same aims was launched, with Galton and his colleagues of the former *Reader* helping.

This new periodical, *Nature*, became the weekly forum of scientists. The first number, under the editorship of Norman Lockyer (1836–1920) appeared in 1869, as did the first volume of the Galtonian Quintet, *Hereditary Genius*. The rest of the quintet were *English Men of Science* (1874c), *Human Faculty* (1883), *Natural Inheritance* (1889) and *Noteworthy Families* (1906). All these were the products of Galton's own educational activities, for, as befitted a referee for *Nature*, he was nothing if not productive. 'Publish or perish' might well have been a maxim of his own, but I attribute their appearance to his born desire, if not to teach, at least to persuade and satisfy his urge to communicate.

Can the influence of the muscular Hughes, who knocked out dockers in the rings at working men's colleges, be seen in Galton's confession of educational faith? For Galton wrote in *Hereditary Genius* (1869):

> I acknowledge freely the great power of education and social influence in developing the active powers of the mind, just as I acknowledge the effect of use in developing the muscles in a blacksmith's arm.

This was in the year before the famous 1870 Education Act was passed, which called into existence locally elected school boards in every town and city in the country.

Thomas Hughes was a more imaginative exegete of the powers of education in public schools, but Galton took a more comprehensive view, embracing both public schools and the new elementary schools, both of which 'helped', as Karl Pearson (1857–1936) wrote, 'to steer Nature's course into smooth waters'. Smooth waters were needed, for in establishing the new schools sectarian rivalries rocked each board election, despite the fact that in their establishment the ballot box (an Australian invention) was used for the first time. And this establishment was Galton's chance, which he took immediately for he believed that the masters in such schools were 'trustworthy and intelligent in no common degree' and that 'they know their pupils well and that the general organisation and discipline of the school is favourable to collecting full and accurate statistics' (1874a).

Correspondence with Alphonse de Candolle (1806–1893) led him to confess that he found de Candolle's conclusions about teaching 'remarkable', and he grudgingly admitted that 'severe teaching sacrifices many original minds but raises the level', adding (this was in 1872, two years after Hughes's brother-in-law, the statesman W. E. Forster (1818–1886) had passed the Education Act) 'we are in the throes of educational reform how best to teach, how best to observe'. The correspondence between these two men – de Candolle presenting the environmental point-of-view and Galton the hereditarian – deserves a study in itself, since de Candolle maintained that Galton 'overstated the influence of heredity' because sociological causes were more important. Galton actually admitted that though deficient in method, de Candolle's book (1872) was 'full of original and suggestive ideas' (Pearson, 1924, vol. 2, pp. 131–56).

Are we then to infer that Galton was influenced by de Candolle's tracing of scientific achievement to education? It certainly stimulated Galton to examine the ancestry and education of English men of science, in the course of which he began to use the term 'correlation' to describe the relationship between smallness of head and energy. He was especially critical of what he called 'educational monopolies' for tempting those with an aptitude for scientific work to uncongenial occupations. He argued that 'professional duties generally ought to be more closely bound up with strictly scientific work than they are at present'.

In 1872 Francis Galton drafted proposals to measure pupils in the board schools, as well as those in the great public schools and in pauper children's schools. His agents for this anthropometry were to be the 'trustworthy and intelligent' masters and teachers in such schools. After it was published in the *Journal of the Anthropological Institute* (1874a), it was followed up by a letter to *Nature* on 5 March 1874 (1874b). As he wrote in *Nature* on 6 May 1880:

> The object of schools should be not only to educate. But also to promote directly and indirectly the science of education. It is astonishing·how little has been done by the school masters of our great public schools in this direction, notwithstanding their enviable opportunities.

Comparing the great British public schools to mediaeval hospitals, where case-taking was unknown, where pathological collections were never dreamt of and where as a consequence 'the art of healing made slow and certain advance', Galton was delighted when the science master of Charterhouse, W. H. Poole, showed the way and commended anthropometry not only to other public schools but to the new board schools. Thus Galton's proposals for obtaining anthropological statistics from schools resulted in anthropological laboratories being established in some schools. The minute beginnings of what is now a vast industry of grant-earning research was pioneered by Galton.

The aggregation of large numbers in board schools, after attendance at schools was made compulsory by the Mundella Act of 1880, threw up the problem of 'fidgeting', since modern activity methods were difficult to apply in schools that were progressively overcrowded. After dutifully measuring what he called 'the rate of fidget', Galton solemnly concluded that it measured the dullness of the performance of the teacher rather than the true mental fatigue of the audience. He opined that boredom in a classroom or audience generally could be detected 'by the unequal horizontal interspace between head and head', as an attentive audience 'all sat upright, their heads almost equidistant'. 'Measure of fidget' was therefore the title of his findings when they were reported in *Nature* on 25 June 1885.

Teachers began to target him as a speaker. In 1883 he was invited to chair a meeting of the educational section of the Teachers' Guild, at which his speech was, not surprisingly, an appeal for help with some of his enquiries. But then, as now, teachers were suspicious of non-teachers conducting research. He received only 116 replies to a widely distributed questionnaire on mental fatigue both in pupils and teachers, though that was an achievement in the days before school secretarial staff.

Well might Karl Pearson, his biographer, remark on 'how little Galton's services to educational reform have been recognised'. As a great supporter of the academic education of women, in 1881 Galton joined 397 others (against 32) in successfully petitioning for the admission of women to examinations at Cambridge.

De Candolle's emphasis on the environmental causation of science proved an incentive to Galton to tease out the qualities of English men of science, and prompted him to write:

> A great and salutary change has undoubtedly come over the feeling of the nation since the time when the leading men of science were boys, for education at that time was conducted in the interests of the clergy and was strongly opposed to science (1874c).

He continued:

> It crushed the enquiring spirit, the love of observation, the pursuit of inductive studies. . . . This gigantic monopoly is yielding, but obstinately and slowly, and it is unlikely that the friends of science will be able, for many years to come, to relax their efforts in educational reform (1874c).

The educative effect of Galton's correspondence with de Candolle can be gauged from a letter Galton wrote to him on 27 December 1872:

> I never said, nor thought, that special aptitudes were inherited so strongly as to be irresistible, which seems to be a dogma you are pleased to ascribe to me and then to repudiate.

It was, indeed, de Candolle's criticism of the hereditarian bias of Galton's work that prompted Galton to write *English Men of Science* (1874c), which was based on his own circularisation and analysis of questionnaires submitted to and filled in by some 180 men of first class scientific status. Their responses convinced him that educational monopolies had caused, as already has been touched on, what he described as 'enormous waste of scientific ability, by inducing those who might have succeeded in science to spend their energies with small effect on uncongenial occupations' (Pearson, 1924, vol. 2, pp. 152–6).

His remedies were not only to extend scientific appointments, assimilating in some cases the English system of teaching into that of the Scots, but also by establishing a system of travelling and other fellowships. The crying evil he detected from his respondents was the 'elaborate machinery for wasting time which has been invented and recommended under the name of "social duties"', the discharge of which, in his opinion absorbed 'much more than half of the progressive force of the nation'. All of this was but a prelude to a five-point curriculum which he deduced from the return of his questionnaire.

By 1891 Galton was being pressurised by Florence Nightingale to appraise the results of the Forster Education Act of twenty-one years before. 'Do we know', she asked Galton,

> (i) what proportion of children forget their whole education after leaving school; whether all they have been taught is waste? . . . the almost accidental statistics of guards recruits would point to a large proportion, (ii) what are the results upon the lives and conduct of children in after life who don't forget all they have been taught? (iii) what are the methods and what are the results, for example in Night Schools and Secondary Schools, in preventing primary education from being a waste? If we know not what are the effects upon our national life of Forster's Act is this not a strange gap on reasonable England's knowledge? (F. Nightingale, 1891).

Galton's reply was uncharacteristically deflating, for as Karl Pearson shrewdly remarked:

> Looking round the possible field of candidates in 1891, what man was there who could have filled the conditions. . . . There was only one man, Galton himself, and it is quite certain that he had not that man in view.

Pearson added:

> We are only just beginning to study social problems – medical, educational, commercial – by adequate statistical methods, and

that study has at present done very little to influence legislation (Pearson, 1924, vol. 2, pp. 418–21).

In 1875 Galton wrote a paper in which he suggested the use of identical twins brought up differently to study the relative values of 'nature and nurture' in human development. In this he put:

> it is to trifling accidental circumstances that the bent of his [man's] disposition and his success are mainly due . . . in fact they do not admit of being tabulated and therefore statistics, however plausible at first sight, are really of very little use (1875).

By 1876 Galton had begun a friendship with Professor George Croom Robertson (1842–1892), Professor of Mental Philosophy and Logic at University College London, who was then starting to edit *Mind*, and whose death sixteen years later, Galton wrote, 'left a void . . . that has never been wholly filled'. Croom Robertson helped Galton with his *Inquiries into Human Faculty and its Development* (1883), in which he tackled the measurement of intelligence and the search for a valid test of ability. And already Galton thought the time was 'near when study of the laws of heredity and of their logical consequences will permeate the philosophy [science] of the university'. For Galton, intelligence was measurable and heritable.

Galton was an educationist of no mean order. Nor am I the only one to argue this, for Dr. R. Ochse in his recent excellent study of the determinants of creative genius has recognised the necessity of a stimulus to prompt genius to attain and maintain excellence (Ochse, 1990).

That stimulus, as far as Galton was concerned, began with mental testing. And one of the landmarks in the development of mental testing was Galton's 1877 'Address to the Anthropological Section of the British Association Meeting', aptly enough at Plymouth which is traditionally associated with adventures to a new world, where he urged anthropologists to turn for a time from physical anthropology to study prevalent types of human character and temperament: psychophysics as it came to be called. He came to the conclusion that early education had a large effect in fixing associations (Pearson, 1924, vol. 2, p. 235).

From a recent biographer of Sir Cyril Burt (1883–1971), L. S. Hearnshaw, we have learnt that Burt's father was Galton's doctor and greatly admired him. It was whilst accompanying his father on his rounds that Cyril Burt first came into contact with Francis Galton, described by Hearnshaw as 'a contact that was a major influence in shaping his career' (Hearnshaw, 1979). As Burt himself acknowledged later in life, his own work at University College was aimed at preserving and developing the Galtonian tradition, 'since in viewpoint and methodology it was essentially Galtonian . . .' Burt's ideas were fundamental to the formulation of the 1944 Education Act and the establishment of selective secondary education based on the testing of intelligence as a constitutional characteristic. As Professor Hearnshaw insists, 'it would be quite wrong . . . to label Burt, just because he advocated the cause of the gifted, as an elitist and hostile to the working class'. He continues:

> He was certainly not biased against the less advantaged members of society. The effect of his work at the London County Council was in fact to reduce the proportion of scholarships going to members of the privileged classes and to double those going to working class children (Hearnshaw, 1979).

True, Burt did entitle his last book *The Gifted Child* (1975), and he did play a great part in the affairs of MENSA, the society founded at Oxford in 1946 by two barristers. But he was knighted (in 1946) under a Labour Government, as – equally fittingly – Galton had been knighted by a Liberal one in 1909.

One hopeful item on the Galtonian agenda however poses an almost insoluble problem to the educationists. As stated in his *Hereditary Genius*, it runs:

> The time may hereafter arrive, in far distant years, when the population of the earth shall be kept strictly within the bounds of number and suitability of race, as the sheep on a well ordered moor or the plants in an orchard house; in the meantime, let us do what we can to encourage the multiplication of the races best fitted to invent and conform to a high and generous civilization, and not, out of a mistaken instinct of giving support to the weak, prevent the incoming of strong and hearty individuals (Galton, 1869, pp. 356–7).

'Some may regard his work as derogatory to education; others will find new hope in his survey of mankind as a highly variable species whose young cannot be thrown into one educational mould', concluded a grand daughter of Charles Darwin, Nora Barlow, in her perceptive assessment (Barlow, 1921) of the relevance of Francis Galton for those who consulted Foster Watson's ambitiously conceived four volume *Encyclopaedia of Education* (1921).

Nora Barlow (1885–1989) stressed Galton's emphasis on the importance of a variety of gifts in the composition of a nation, his insistence on the numerical representation of data, and his deprecation of any pursuit of a uniform type. For Galton held that 'men differ in a variety of ways almost as profoundly as animals in different cages in the zoological gardens . . . each may be good of its type'.

The *Encyclopaedia of Education*, published on the crest of the wave of idealism after the First World War, professed to offer a 'comprehensive, practical and authoritative guide on matters connected with education, including educational principles and practice, various types of teaching instruction and educational systems throughout the world'. Hence it contained a full entry by Leonard Darwin (1850–1943) on the Eugenics Education Society, which it described as 'founded in 1907 to carry out the following objects':

1. Persistently to set forth the national importance of eugenics in order to modify public opinion and to create a sense of responsibility in respect of bringing matters pertaining to human parenthood under the domination of eugenic ideals.

2. To spread a knowledge of the laws of heredity as far as they are surely known, and so far as that knowledge might affect the improvement of race.

3. To further eugenic teaching at home, in the schools and elsewhere (L. Darwin, 1921).

Citing the achievement of the society in promoting the Mental Deficiency Act of 1913 as 'the only piece of English social legislation extant in which the influence of heredity has been treated as a practical factor in determining its provisions', the Eugenics Education Society was further credited with urging local authorities to secure the effective working of the act. It also credited the society with having advocated 'for a long time the advisability of giving far more substantial allowances for children in the assessment for income tax' in order to encourage worthy parenthood. In addition it attributed the steps taken by the framers of the 1920 budget as being a consequence of what it called 'this agitation' (L. Darwin, 1921).

Writing in *Inquiries into Human Faculty* (1883, p. 208), Galton's view of a boy was that he was 'born prepared to attach himself as a climbing plant is naturally disposed to climb, the kind of stick being of little importance'. This could be regarded as an anticipation of a comprehensive support.

Francis Galton was Honorary President of the Eugenics Education Society, founded in 1907, from 1908 until his death in 1911, when Leonard Darwin became president until 1929. The Eugenics Education Society became the Eugenics Society in 1926, which changed its name to the Galton Institute in 1989. The dropping of the word 'education' – a word rightly used in my opinion – from the title of the society in 1926, together with the advocacy by at least one British biologist that sterilisation be applied as a punishment for parents who had to resort to public assistance in order to support their children, before the Second World War, somewhat dimmed the image of the society and not even its other activities could offset this. But the adhesion of radicals like Professor J. D. Bernal (1901–1971), in wartime, did something to revive it.

As a conclusion, one could not improve on a quotation from the draft of the scheme for the 'Francis Galton Laboratory for the Study of National Eugenics' presented to the University of London by Karl Pearson (and written in conjunction with Francis Galton) in 1907. This defined 'national eugenics' as 'the study of the agencies under social control that may improve or impair the racial qualities of future generations either physically or mentally', and indicated that the agencies were to include schools and colleges (Pearson, 1930, vol. 3A, pp. 304–7).

References

Barlow, Nora (1921) 'Francis Galton', in Foster Watson, *The Encyclopaedia and Dictionary of Education* (London: Pitman) vol. 2, pp. 659–60.
Burt, Cyril (1975) *The Gifted Child* (London: Hodder and Stoughton).
Darwin, Leonard (1921) 'Eugenics at School', in Foster Watson, *The Encyclopaedia and Dictionary of Education* (London: Pitman) vol. 2, pp. 586–8.
de Candolle, Alphonse (1872) *Histoire des Sciences et des Savants depuis deux Siècles*.

Galton, F. (1869) *Hereditary Genius* (London: Macmillan).

Galton, F. (1872) letter in Karl Pearson (1924) *The Life, Letters and Labours of Francis Galton* (Cambridge at the University Press) vol. 2, pp. 135–6.

Galton, F. (1874a) 'Proposal to apply for anthropological statistics from schools', *Journal of the Anthropological Institute*, vol. 3, pp. 308–11.

Galton, F. (1874b) 'On a proposed statistical scale', letter in *Nature*, vol. 9, pp. 342–3.

Galton, F. (1874c) *English Men of Science: their Nature and Nurture* (London: Macmillan).

Galton, F. (1875) 'The History of Twins, as a criterion of the relative powers of nature and nurture', *Fraser's Magazine*, vol. 12, pp. 566–76. Revised version reprinted in *Journal of the Anthropological Institute*, vol. 5, pp. 391–406; summary in *Nature*, vol. 13, p. 59.

Galton, F. (1877) *Address to the Department of Anthropology, Section H*, British Association Report, pp. 94–100. Summary in *Nature*, vol. 16, pp. 344–7. Also under the title *Address to the Anthropological Department of the British Association* (London: W. Clowes & Sons).

Galton, F. (1880) 'The Opportunities of Science Masters at Schools', letter in *Nature*, vol. 22, pp. 9–10.

Galton, F. (1883) *Inquiries into Human Faculty and its Development* (London: Macmillan).

Galton, F. (1885) 'The Measure of Fidget', *Nature*, vol. 32, pp. 174–5.

Galton, F. (1889) *Natural Inheritance* (London: Macmillan).

Galton, F. with E. Schuster (1906) *Noteworthy Families* (London: John Murray).

Hearnshaw, L. S. (1979) *Cyril Burt, Psychologist* (Ithaca: Cornell University Press).

Nightingale, Florence (1891) letter in Karl Pearson (1924) *The Life, Letters and Labours of Francis Galton* (Cambridge at the University Press) vol. 2, pp. 416–18.

Ochse, R. (1990) *Before the Gates of Excellence* (Cambridge University Press).

Pearson, Karl (1914, 1924 and 1930) *The Life, Letters and Labours of Francis Galton*, 3 vols. (Cambridge at the University Press).

Watson, Foster (ed.) (1921) *The Encyclopaedia and Dictionary of Education*, 4 vols (London: Sir Isaac Pitman & Sons).

14 The Galton Laboratory, University College London

J. S. Jones

Francis Galton opened his first Anthropometric Laboratory at the International Health Exhibition of 1884. Here people were measured at a charge of 3d each and given a copy of their measurements, while the schedule was kept for his records. According to a contemporary newspaper cutting, nearly 10 000 people were measured; we still have some of these records in the department. In 1888 he set up a more permanent laboratory, also called the Anthropometric Laboratory, built at his own expense on land given by the commissioners of the 1851 exhibition and reached through the Western Exhibition Galleries at the Science Museum in South Kensington. Later the land became the property of the Imperial Institute who wanted it for building; the museum then offered him space in their own building, where the second Anthropometric Laboratory opened in 1891.

The connection with University College appears to have come about through his acquaintance with W. F. R. Weldon (1860–1906), Jodrell Professor of Zoology and Comparative Anatomy. They both served on the Royal Society Committee for the Measurement of Plants and Animals, later renamed Evolution (Plants & Animals) Committee, and in 1896 Karl Pearson (1857–1936) also joined this committee. All three left the committee in 1900 when William Bateson (1861–1926) and his supporters 'captured' it. In 1894 Galton gave the first lectures in variation and correlation at University College.

In 1904 Galton founded and directed the Eugenics Record Office. The accommodation at 88 Gower Street was provided by the University of London, and he offered a grant of £500 a year for three years to endow a fellowship in national eugenics within the university, which was accepted in October 1904. He wanted it to be associated with the Drapers' Company Biometric Laboratory, which was established in the same year, with Karl Pearson as director, with a grant from the Drapers' Company. Edgar Schuster was the first

Francis Galton Research Fellow. In 1906 Pearson took over the directorship from Galton, who was then in his eighties.

In 1907 a draft scheme for a change in name to the 'Francis Galton Laboratory for the Study of National Eugenics' was presented to the University by Pearson. In this it was stated that the existing laboratory was under the supervision of Pearson in consultation with Galton. The proposed staff for 1907 were David Heron as Fellow (succeeding E. Schuster) Ethel M. Elderton as Francis Galton Scholar and Amy Barrington as computer. This draft was accepted. It was also decided that the Galton Fellowship Committee, appointed in 1904, should be renamed the Francis Galton Laboratory Committee.

From the start there seems to have been a good deal of inconsistency in the titles given to the laboratory. 'The Study of' seems to have been soon dropped from it, and the name 'The Francis Galton Laboratory for National Eugenics' appears on all the *Eugenics Laboratory Memoirs* from 1907 till 1925. Inside the covers the advertisement for other publications of the laboratory is headed 'The Francis Galton Eugenics Laboratory'. Galton and Pearson, in their correspondence, referred to it familiarly as the Eugenics Laboratory right from the start, and the laboratory's memoirs, from when they first appeared in 1907, were always called the *Eugenics Laboratory Memoirs*. In 1907 the laboratory was transferred from 88 Gower Street to the south wing at University College, where it remained until 1920. Galton gave a further £1000 to continue the £500 grant for 1908 and 1909.

History does not relate how the laboratory survived in 1910 when we have no record of any grant. In 1911 Sir Francis Galton died and left his effects and all the residue of his estate to the University of London to found a professorship, to be called the Galton Professorship of Eugenics, expressing the wish that Karl Pearson should be the first professor and that he should be enabled to keep his directorship of the Biometric Laboratory.

A trust for the Galton Chair was set up, to be administered by University College on behalf of the university, and the professorship was established with Pearson as first professor in 1911. Pearson called the department the Department of Applied Statistics, which was to include both the Drapers' Company Biometric Laboratory and the Francis Galton Laboratory for National Eugenics – on his notepaper the title had been shortened to Francis Galton Eugenics Laboratory by 1923. It occupied much of the first floor of the south side of the college quadrangle. The university had intended to erect a new building to house the Galton Laboratory, but Galton's legacy turned out to be much less than had at first been supposed. In fact a solicitor's clerk had by mistake added an extra figure to the sum declared for probate, and when this was discovered probate had to be sought again for the reduced amount. As there was not enough money for the plans proposed it was decided to appeal to the public, and in 1911 the senate launched an appeal for £15 000 for a building, the University providing the site.

In 1912 Sir Herbert Bartlett offered a building and work was begun. But it was not occupied by the laboratory until 1920 as war intervened, and from 1914 to 1919 it was used as a military hospital. In October 1919 departmental work began there, and the Bartlett Building was officially opened on 4 June 1920.

In 1913 the Senate passed a resolution that the Department of Applied Statistics should in future be the Department of Applied Statistics and Eugenics, and should include the Biometric and Galton Laboratories. The Francis Galton Committee was to be replaced by a Galton Subcommittee to be appointed yearly. Dr. Heron was re-appointed as assistant director, Miss Elderton as Galton Research Fellow, and Amy Barrington as research assistant.

The Annals of Eugenics was launched in 1925 by Pearson, assisted by Miss Elderton. It was subtitled 'A journal for the scientific study of racial problems', and was issued by the Francis Galton Laboratory for National Eugenics and printed by the Cambridge University Press.

In 1933 Karl Pearson retired. In a report by the professorial board of University College arising from this (10 April 1933, Appendix VIII), it was decided that the Departments of Statistics and Eugenics should be separated – eugenics to include anthropometry – and that the staff of the present department should be distributed between the two. The Galton Professor was to be head of the Department of Eugenics, and the Department of Statistics was to be headed by a reader or another professor. It was recommended that the Galton collections should be in the hands of the Galton Subcommittee, and that the curator, who should be the Galton Professor, should be responsible to the subcommittee.

R. A. Fisher (1890–1962) succeeded Karl Pearson as Galton Professor and edited the *Annals of Eugenics* from volume VI (1934–5) onward, with an editorial committee consisting of R. Ruggles Gates, J. B. S. Haldane, M. N. Karn, J. A. Fraser Roberts and Lord

Horder. It was sub-titled 'A journal devoted to the genetic study of human population' and was published by the Galton Laboratory for National Eugenics and the Eugenics Society. This arrangement presumably was continued until the outbreak of war, for volume X (1940) still bore this imprint, but volume XI (1941–2) omits Lord Horder and the Eugenics Society from the title page.

In 1936 Pearson died, and Mrs. Weldon, W. F. R. Weldon's widow, died later in the same year, leaving a sum of money to found a chair of biometry, preferably at University College. In the will, made many years before she died, the hope was expressed that the Biometry Professor would work with the Galton Professor. It was finally decided the chair would be attached to zoology and offered to J. B. S. Haldane (1892–1964).

In 1944 L. S. Penrose (1898–1972) succeeded R. A. Fisher as Galton Professor, and biometry and eugenics were combined into one department headed by J. B. S. Haldane. Penrose also became the editor of the *Annals of Eugenics* from volume 13 (1946–7) onwards, with the assistance of J. Bell, R. A. Fisher, J. B. S. Haldane, M. N. Karn, J. Fraser Roberts and R. R. Race. It was now subtitled 'A journal of human genetics' and was published by the Galton Laboratory, University College. In 1950 an agreement was made with Cambridge University Press that they should publish the *Annals* for the Laboratory, beginning with volume 16 (1951–2). In 1954 the title was changed to the *Annals of Human Genetics*, first used for volume 19, which had the old title (*Annals of Eugenics*) printed below the new one with no other sub-title.

Haldane retired from the Weldon Chair of Biometry in 1957, and in 1958 Penrose, as Galton Professor, became head of the Department of Eugenics, Biometry and Genetics. Penrose retired in 1965, and Harry Harris became head of department – now renamed the Department of Human Genetics and Biometry – as Galton Professor and Director of the Galton Laboratory. C. A. B. Smith was appointed Weldon Professor of Biometry in 1964 and became Emeritus Professor in 1982, since which time the chair has been in abeyance. Harry Harris and C. A. B. Smith became joint editors of the *Annals of Human Genetics* (volume 30) in 1965, when the old title *Annals of Eugenics* was dropped from the title page. The journal had been edited with the assistance of C. A. B. Smith since volume 20 (1955–6), and R. A. Fisher ceased to be associated with it after volume 25 (1961–2). When Harris and Smith took over the editorship, those assisting them were Julia Bell, Jean Edmiston, Mary

N. Karn, L. S. Penrose, R. R. Race and J. A. Fraser Roberts.

In September 1967 the Department of Human Genetics and Biometry and the Galton Laboratory were moved to Wolfson House, built with a gift from Sir Isaac Wolfson in a back street close to the main campus of University College. Harry Harris resigned in 1976, and E. Bette Robson became Galton Professor, director of the Galton Laboratory and head of the Department of Genetics and Biometry in 1978. She was succeeded by J. S. Jones as head of the Department of Genetics and Biometry at University College London in 1990.

The Galton Laboratory is the lineal descendant of the Anthropometric Laboratory set up by Francis Galton at the International Health Exhibition of 1884, and of the Eugenics Record Office, also started by him, and the Biometric Laboratory, each founded at University College in 1904. Since then it has maintained an unbroken association with UCL, and now contains within itself the academic staff of UCL Department of Genetics and Biometry, as well as a group of Medical Research Council-funded staff associated with research and teaching in genetics and amounting to more than fifty scientists. It continues to be concerned with the publication of the *Annals of Human Genetics* (originally the *Annals of Eugenics*), the oldest of the scientific journals concerned with this subject, and contains many of the relics of Galton himself, including his scientific library with his autographed copy of *The Origin of Species*, the Quincunx (Galton's ingenious machine for modelling statistical distributions), as well as other, perhaps curious, possessions of its founder.

Galton was the founder of human genetics, and after a long period of relative stagnation, largely due to the practical difficulties of using standard genetical approaches on people, this subject has become the fastest growing area of genetics, if not of biology as a whole. It has followed a technical revolution in the study of genes, so that we now know as much about the structure of human genomes and the genetics of human evolution as we do about these aspects of the biology of any other organism. At present the four main areas of research at the Galton Laboratory are human biochemical and molecular genetics, diagnostic cytogenetics and cancer, developmental genetics and evolutionary genetics.

15 Three Memoirs

Sir Francis Galton

(*Note*: The first edition of Sir Francis Galton's *Inquiries into Human Faculty* was published in 1883 by Macmillan & Co., but in the second edition, published in 1907 in their Everyman's Library by J. M. Dent & Sons, three chapters – *Enthusiasm, Theocratic Intervention* and *Objective Efficacy of Prayer* – were excised. These are now reprinted here, but the third, which had originally been a separate paper, 'Statistical inquiries into the efficacy of prayer' in the *Fortnightly Review* (vol. 12, pp. 125–35) in 1872, had been abbreviated in *Inquiries* and now appears in its original form. About the abbreviation, Francis Galton later wrote (in the preface of the second edition of *Inquiries* in 1907) that it had given 'a somewhat inexact impression of its object, which was to investigate certain views then thought orthodox, but which are growing obsolete').

STATISTICAL INQUIRIES INTO THE EFFICACY OF PRAYER

An eminent authority has recently published a challenge to test the efficacy of prayer by actual experiment. I have been induced, through reading this, to prepare the following memoir for publication, nearly the whole of which I wrote and laid by many years ago, after completing a large collection of data, which I had undertaken for the satisfaction of my own conscience.

The efficacy of prayer seems to me a simple, as it is a perfectly appropriate and legitimate subject of scientific inquiry. Whether prayer is efficacious or not, in any given sense, is a matter of fact on which each man must form an opinion for himself. His decision will be based upon data more or less justly handled, according to his education and habits. An unscientific reasoner will be guided by a confused recollection of crude experience. A scientific reasoner will scrutinize each separate experience before he admits it as evidence, and will compare all the cases he has selected on a methodical system.

The doctrine commonly preached by the clergy is well expressed in the most recent, and by far the most temperate and learned of

theological encyclopædias, namely, *Smith's Dictionary of the Bible.*
The article on 'Prayer,' written by the Rev. Dr. Barry, states as
follows: 'Its real objective efficacy . . . is both implied and expressed
(in Scripture) in the plainest terms. . . . We are encouraged to ask
special blessings, both spiritual and temporal, in hopes that thus, and
thus only, we may obtain them. . . . It would seem the intention of
Holy Scripture to encourage all prayer, more especially intercession,
in all relations and for all righteous objects.' Dr. Hook, the present
Dean of Chichester, states in his *Church Dictionary*, under 'Prayer,'
that 'the general providence of God acts through what are called the
laws of nature. By this particular providence God interferes with those
laws, and he has promised to interfere in behalf of those who pray in the
name of Jesus. . . . We may take it as a general rule that we may pray
for that for which we may lawfully labour, and for that only.'

The phrases of our Church service amply countenance this view;
and if we look to the practice of the opposed sections of the religious
world, we find them consistent in maintaining it. The so-called 'Low
Church' notoriously places absolute belief in special providences
accorded to pious prayer. This is testified by the biographies of its
members, the journals of its missionaries, and the 'united prayer
meetings' of the present day. The Roman Catholics offer religious
vows to avert danger; they make pilgrimages to shrines; they hang
votive offerings and pictorial representations, sometimes by thousands, in their churches, of fatal accidents averted by the manifest interference of a solicited saint.

A *prima facie* argument in favour of the efficacy of prayer is
therefore to be drawn from the very general use of it. The greater
part of mankind, during all the historic ages, have been accustomed
to pray for temporal advantages. How vain, it may be urged, must be
the reasoning that ventures to oppose this mighty consensus of belief!
Not so. The argument of universality either proves too much, or else
it is suicidal. It either compels us to admit that the prayers of Pagans,
of Fetish worshippers, and of Buddhists who turn praying wheels, are
recompensed in the same way as those of orthodox believers; or else
the general consensus proves that it has no better foundation than the
universal tendency of man to gross credulity.

The collapse of the argument of universality leaves us solely concerned with a simple statistical question – are prayers answered, or
are they not? There are two lines of research, by either of which we
may pursue this inquiry. The one that promises the most trustworthy
results is to examine large classes of cases, and to be guided by broad

averages; the other, which I will not employ in these pages, is to deal with isolated instances. An author who made much use of the latter method might reasonably suspect his own judgment – he would certainly run the risk of being suspected by others – in choosing one-sided examples.

The principles are broad and simple upon which our inquiry into the efficacy of prayer must be established. We must gather cases for statistical comparison, in which the same object is keenly pursued by two classes similar in their physical but opposite in their spiritual state; the one class being prayerful, the other materialistic. Prudent pious people must be compared with prudent materialistic people, and not with the imprudent nor the vicious. Secondly, we have no regard, in this inquiry, to the course by which the answer to prayers may be supposed to operate. We simply look to the final result – whether those who pray attain their objects more frequently than those who do not pray, but who live in all other respects under similar conditions. Let us now apply our principles to different cases.

A rapid recovery from disease may be conceived to depend on many causes besides the reparative power of the patient's constitution. A miraculous quelling of the disease may be one of these causes; another is the skill of the physician, or of the nurse; another is the care that the patient takes of himself. In our inquiry, whether prayerful people recover more rapidly than others under similar circumstances, we need not complicate the question by endeavouring to learn the channel through which the patient's prayer may have reached its fulfilment. It is foreign to our present purpose to ask if there be any signs of a miraculous quelling of the disease, or if, through the grace of God, the physician had showed unusual wisdom, or the nurse or the patient unusual discretion. We simply look to the main issue – do sick persons who pray, or are prayed for, recover on the average more rapidly than others?

It appears that, in all countries and in all creeds, the priests urge the patient to pray for his own recovery, and the patient's friends to aid him with their prayers; but that the doctors make no account whatever of their spiritual agencies, unless the office of priest and medical man be combined in the same individual. The medical works of modern Europe teem with records of individual illnesses and of broad averages of disease, but I have been able to discover hardly any instance in which a medical man of any repute has attributed recovery to the influence of prayer. There is not a single instance, to my knowledge, in which papers read before statistical societies have recognized the agency of prayer either on disease or on anything else. The universal habit of the scientific world to ignore the agency of prayer is a very important fact. To fully appreciate the 'eloquence of the silence' of medical men, we must bear in mind the care with which they endeavour to assign a sanatory value to every influence. Had prayers for the sick any notable effect, it is incredible but that the doctors, who are always on the watch for such things, should have observed it, and added their influence to that of the priests towards obtaining them for every sick man. If they abstain from doing so, it is not because their attention has never been awakened to the possible efficacy of prayer, but, on the contrary, that although they have heard it insisted on from childhood upwards, they are unable to detect its influence. Most people have some general belief in the objective efficacy of prayer, but none seem willing to admit its action in those special cases of which they have scientific cognizance.

Those who may wish to pursue these inquiries upon the effect of prayers for the restoration of health could obtain abundant materials from hospital cases, and in a different way from that proposed in the challenge to which I referred at the beginning of these pages. There are many common maladies whose course is so thoroughly well understood as to admit of accurate tables of probability being constructed for their duration and result. Such are fractures and amputations. Now it would be perfectly practicable to select out of the patients at different hospitals under treatment for fractures and amputations two considerable groups; the one consisting of markedly religious and piously befriended individuals, the other of those who were remarkably cold-hearted and neglected. An honest comparison of their respective periods of treatment and the results would manifest a distinct proof of the efficacy of prayer, if it existed to even a minute fraction of the amount that religious teachers exhort us to believe.

An inquiry of a somewhat similar nature may be made into the longevity of persons whose lives are prayed for; also that of the praying classes generally; and in both these cases we can easily obtain statistical facts. The public prayer for the sovereign of every state, Protestant and Catholic, is and has been in the spirit of our own, 'Grant her in health long to live.' Now, as a simple matter of fact, has this prayer any efficacy? There is a memoir by Dr. Guy, in the *Journal of the Statistical Society* (vol. xxii, p. 355), in which he compares the mean age of sovereigns with that of other classes of persons. His results are expressed in the following table:–

MEAN AGE ATTAINED BY MALES OF VARIOUS CLASSES WHO HAD SURVIVED THEIR 30TH YEAR, from 1758 to 1843. Deaths by accident or violence are excluded

	Number	Average	Eminent Men *
Members of Royal Houses	97	64.04	—
Clergy	945	69.49	66.42
Lawyers	294	68.14	66.51
Medical profession	244	67.31	67.07
English aristocracy	1,179	67.31	—
Gentry	1,632	70.22	—
Trade and commerce	513	68.74	—
Officers in the Royal Navy	366	68.40	—
English literature and science	395	67.55	65.22
Officers of the Army	569	67.07	—
Fine Arts	239	65.96	64.74

* The eminent men are those whose lives are recorded in *Chalmer's Biography*, with some additions from the *Annual Register*.

The sovereigns are literally the shortest lived of all who have the advantage of affluence. The prayer has therefore no efficacy, unless the very questionable hypothesis be raised, that the conditions of royal life may naturally be yet more fatal, and that their influence is partly, though incompletely, neutralized by the effects of public prayers.

It will be seen that the same table collates the longevity of clergy, lawyers, and medical men. We are justified in considering the clergy to be a far more prayerful class than either of the other two. It is their profession to pray, and they have the practice of offering morning and evening family prayers in addition to their private devotions. A reference to any of the numerous published collections of family prayers will show that they are full of petitions for temporal benefits. We do not, however, find that the clergy are in any way more long lived in consequence. It is true that the clergy, as a whole show a life-value of 69.49, as against 68.14 for the lawyers, and 67.31 for the medical men; but the easy country life and family repose of so many of the clergy are obvious sanatory conditions in their favour. This difference is reversed when the comparison is made between distinguished members of the three classes – that is to say, between persons of sufficient note to have had their lives recorded in a biographical dictionary. When we examine this category, the value of life among the clergy, lawyers, and medical men is as 66.42, 66.51, and 67.07 respectively, the clergy being the shortest lived of the three. Hence the prayers of the clergy for protection against the perils and dangers of the night, for protection during the day, and for recovery from sickness, appear to be futile in result.

In my work on *Hereditary Genius*, and in the chapter on 'Divines,' I have worked out the subject with some minuteness on other data, but with precisely the same result. I show that the divines are not specially favoured in those worldly matters for which they naturally pray, but rather the contrary, a fact which I ascribe in part to their having, as a class, indifferent constitutional vigour. I give abundant reason for all this, and do not care to repeat myself; but I should be glad if such of the readers of this present paper who may be accustomed to statistics would refer to the chapter I have mentioned. They will find it of use in confirming what I say here. They will believe me the more when I say that I have taken considerable pains to get at the truth in the questions raised in this present memoir, and that when I was engaged upon them, I worked, so far as my material went, with as much care as I gave to that chapter on 'Divines'; and lastly, they will understand that, when writing the chapter in question, I had all this material by me unused, which justified me in speaking out as decidedly as I did then.

A further inquiry may be made into the duration of life among missionaries. We should lay greater stress upon their mortality than upon that of the clergy, because the laudable object of a missionary's career is rendered almost nugatory by his early death. A man goes, say to a tropical climate, in the prime of manhood, who has the probability of many years of useful life before him, had he remained at home. He has the certainty of being able to accomplish sterling good as a missionary, if he should live long enough to learn the language and habits of the country. In the interval he is almost useless. Yet the painful experience of many years shows only too clearly that the missionary is not supernaturally endowed with health. He does not live longer than other people. One missionary after another dies shortly after his arrival. The work that lay almost within the grasp of each of them lingers incompleted.

It must here be repeated, that comparative immunity from disease compels the suspension of no purely material law, if such an expression be permitted. Tropical fever, for example, is due to many subtle causes which are partly under man's control. A single hour's exposure to sun, or wet, or fatigue, or mental agitation, will determine an attack. Now even if God acted only on the minds of the missionaries, his action might be as much to the advantage of their health as if he wrought a physical miracle. He could disincline them to take those courses which might result in mischance, such as the forced march, the wetting, the abstinence from food, or the night exposure, any one of which was competent to develop the fever that struck them down. We must not dwell upon the circumstances of individual cases, and say 'this was a providential escape,' or 'that was a salutary chastisement,' but we must take the broad averages of mortality, and, when we do so, we find that the missionaries do not form a favoured class.

The efficacy of prayer may yet further be tested by inquiry into the proportion of deaths at the time of birth among the children of the praying and the non-praying classes. The solicitude of parents is so powerfully directed towards the safety of their expected offspring as to leave no room to doubt that pious parents pray fervently for it, especially as death before baptism is considered a most serious evil by many Christians. However, the distribution of still-births appears wholly unaffected by piety. The proportion, for instance, of the still-births published in the *Record* newspaper and in the *Times* was found by me, on an examination of a particular period, to bear an identical relation to the total number of deaths. This inquiry might easily be pursued by those who consider that more ample evidence was required.

When we pray in our Liturgy 'that the Nobility may be endued with grace, wisdom and understanding,' we pray for that which is clearly incompatible with insanity. Does that frightful scourge spare our nobility? Does it spare very religious people more than others? The answer is an emphatic negative to both of these questions. The nobility, probably from their want of the wholesome restraints felt in humbler walks of life, and from their intermarriages, and the very religious people of all denominations, probably from their meditations on hell, are peculiarly subject to it. Religious madness is very common indeed.

As I have already hinted, I do not propose any special inquiry whether the general laws of physical nature are ever suspended in fulfilment of prayer: whether, for instance, success has attended the occasional prayers in the Liturgy when they have been used for rain, for fair weather, for the stilling of the sea in a storm, or for the abatement of a pestilence. I abstain from doing so for two reasons.

First, if it is proved that God does not answer one large class of prayers at all, it would be of less importance to pursue the inquiry. Secondly, the modern feeling of this country is so opposed to a belief in the occasional suspension of the general laws of nature, that an English reader would merely smile at such an investigation.

If we are satisfied that the actions of man are not influenced by prayer, even through the subtle influences of his thoughts and will, the only probable form of agency will have been disproved, and no one would care to advance a claim in favour of direct physical interferences.

Biographies do not show that devotional influences have clustered in any remarkable degree round the youth of those who, whether by their talents or social position, have left a mark upon our English history. Lord Campbell, in his preface to his *Lives of the Chancellors*, says, 'There is no office in the history of any nation that has been filled with such a long succession of distinguished and interesting men as the office of Lord Chancellor,' and that 'generally speaking, the most eminent men, if not the most virtuous, have been selected to adorn it.' His implied disparagement of their piety is fully sustained by an examination of their respective biographies, and by a taunt of Horace Walpole, quoted in the same preface. An equal absence of remarkable devotional tendencies may be observed in the lives of the leaders of great political parties. The founders of our great families too often owed their advancement to tricky and time-serving court-iership. The belief so frequently expressed in the Psalms, that the descendants of the righteous shall continue, and that those of the wicked shall surely fail, is not fulfilled in the history of our English peerage. Take for instance the highest class, that of the Ducal houses. The influence of social position in this country is so enormous that the possession of a dukedom is a power that can hardly be understood without some sort of calculation. There are, I believe, only twenty-seven dukes to about eight millions of adult male Englishmen, or about three dukes to each million, yet the cabinet of fourteen ministers which governs this country, and India too, commonly contains one duke, often two, and in recent times three. The political privilege inherited with a dukedom in this country is at the lowest estimate many thousand-fold above the average birth-right of Englishmen. What was the origin of these ducal families whose influence on the destiny of England and her dependencies is so enormous? Were their founders the eminently devout children of eminently pious parents? Have they and their ancestors been distinguished among the praying classes? Not so. I give in a footnote * a list of their names, which recalls many a deed of patriotism, valour, and skill, many an instance of eminent merit of the worldly sort, which we Englishmen honour six days out of the seven – many scandals, many a disgrace, but not, on the other hand, a single instance known to me of eminently prayerful qualities. Four at least of the existing ducal houses are unable to claim the title of having been raised into existence through the devout habits of their progenitors, because the families of Buccleuch, Grafton, St. Albans, and Richmond were thus highly ennobled solely on the ground of their being descended from Charles II and four of his mistresses, namely, Lucy Walters, Barbara Villiers, Nell Gwynne, and Louise de Querouaille. The dukedom of Cleveland may almost be reckoned as a fifth instance.

The civil liberty we enjoy in England, and the energy of our race, have given rise to a number of institutions, societies, commercial adventures, political meetings, and combinations of all sorts. Some of these are exclusively clerical, some lay, and others mixed. It is impossible for a person to have taken an active share in social life without having had abundant means of estimating for himself, and of hearing the opinion of others, on the value of a preponderating clerical element in business committees. For my own part, I never heard a favourable one. The procedure of Convocation, which, like all exclusively clerical meetings, is opened with prayer, has not inspired the outer world with much respect. The histories of the great councils of the Church are most painful to read. There is reason to expect that devout and superstitious men should be unreasonable; for a person who believes his thoughts to be inspired, necessarily accredits his prejudices with divine authority. He is therefore little accessible to argument, and he is intolerant of those whose opinions differ from his, especially on first principles. Consequently he is a bad coadjutor in business matters. It is a common weekday opinion of the world that praying people are not practical.

Again, there is a large class of instances where an enterprise on behalf of pious people is executed by the agency of the profane. Do such enterprises prosper beyond the average? For instance, a vessel on a missionary errand is navigated by ordinary seamen. A fleet,

---

* Abercorn, Argyll, Athole, Beaufort, Bedford, Buccleuch, Buckingham, Cleveland, Devonshire, Grafton, Hamilton, Leeds, Leinster, Manchester, Marlborough, Montrose, Newcastle, Norfolk, Northumberland, Portland, Richmond, Roxburghe, Rutland, St. Albans, Somerset, Sutherland, Wellington.

followed by the prayers of the English nation, carries reinforcements to quell an Indian mutiny. We do not care to ask whether the result of these prayers is to obtain favourable winds, but simply whether they ensue in a propitious voyage, whatever may be the agencies by which that result was obtained. The success of voyages might be due to many other agencies than the suspension of the physical laws that control the winds and currents; just as we showed that a rapid recovery from illness might be due to other causes than direct interference with cosmic order. It might have been put into the captain's heart to navigate in that course and to perform those acts of seamanship which proved links in a chain that led to eventual success. A very small matter would suffice to make a great difference in the end. A vessel navigated by a man who was a good forecaster of weather and an accomplished hydrographer would considerably outstrip another that was deficient in so accomplished a commander, but otherwise similarly equipped. The perfectly instructed navigator would deviate from the most direct course by perhaps some mere trifle, first here, then there, in order to bring his vessel within favouring slants of wind and advantageous currents. A ship commanded by a captain and steered by a sailor whose hearts were miraculously acted upon in answer to prayer would unconsciously, as by instinct, or even as it were by mistake, perform these deviations from routine, which would lead to ultimate success.

The missionaries who are the most earnestly prayed for are usually those who sail on routes where there is little traffic, and therefore where there is more opportunity for the effects of secret providential overruling to display themselves than among those who sail in ordinary sea voyages. In the usual sea routes a great deal is known of the peculiarities of the seasons and currents, and of the whereabouts of hidden dangers of all kinds; their average risk is small, and the insurance is low. But when vessels are bound to ports like those sought by the missionaries the case is different. The risk that attends their voyages is largely increased, and the insurance is proportionately raised. But is the risk equally increased in respect to missionary vessels and to those of traders and slave-dealers? The comparison between the fortune that attends prayerful and non-prayerful people may here be most happily made. The missionaries are eminently among the former category, and the slave-dealers and traders we speak of in the other. Traders in the unhealthy and barbarous regions to which we refer are notoriously the most godless and reckless (on the broad average) of any of their set. We have, unfortunately, little knowledge of the sea risks of slavers, because the rates of their insurance involve the risk of capture. There is, however, a universal testimony, in the parliamentary reports on slavery, to the excellent and skilful manner in which these vessels are sailed and navigated, which is a *prima facie* reason for believing their sea risks to be small. As to the relative risks run by ordinary traders and missionary vessels, the insurance offices absolutely ignore the slightest difference between them. They look to the class of the vessel, and to the station to which she is bound, and to nothing else. The notion that a missionary or other pious enterprise carries any immunity from danger has never been entertained by insurance companies.

To proceed with our inquiry, whether enterprises on behalf of pious people succeed better than others when they are intrusted to profane hands, we may ask – Is a bank or other commercial undertaking more secure when devout men are among its shareholders – or when the funds of pious people, or charities, or of religious bodies are deposited in its keeping, or when its proceedings are opened with prayer, as was the case with the disastrous Royal British Bank? It is impossible to say yes. There are far too many sad experiences of the contrary.

If prayerful habits had influence on temporal success, it is very probable, as we must again repeat, that insurance offices, of at least some descriptions, would long ago have discovered and made allowance for it. It would be most unwise, from a business point of view, to allow the devout, supposing their greater longevity even probable, to obtain annuities at the same low rates as the profane. Before insurance offices accept a life, they make confidential inquiries into the antecedents of the applicant. But such a question has never been heard of as, 'Does he habitually use family prayers and private devotions?' Insurance offices, so wakeful to sanatory influences, absolutely ignore prayer as one of them. The same is true for insurances of all descriptions, as those connected with fire, ships, lightning, hail, accidental death and cattle sickness. How is it possible to explain why Quakers, who are most devout and most shrewd men of business, have ignored these considerations, except on the ground that they do not really believe in what they and others freely assert about the efficacy of prayer? It was at one time considered an act of mistrust in an over-ruling Providence to put lightning conductors on churches; for it was said that God would surely take care of his own. But Arago's collection of the accidents from lightning showed they were sorely needed; and now lightning conductors are universal.

Other kinds of accidents befall churches, equally with other buildings of the same class; such as architectural flaws, resulting in great expenses for repair, fires, earthquakes, and avalanches.

The cogency of all these arguments is materially increased by the recollection that many items of ancient faith have been successively abandoned by the Christian world to the domain of recognized superstition. It is not two centuries ago, long subsequent to the days of Shakespeare and other great names, that the sovereign of this country was accustomed to lay hands on the sick for their recovery, under the sanction of a regular Church service, which was not omitted from our prayer-books till the time of George II. Witches were unanimously believed in, and were regularly exorcised, and punished by law, up to the beginning of the last century. Ordeals and duels, most reasonable solutions of complicated difficulties according to the popular theory of religion, were found absolutely fallacious in practice. The miraculous power of relics and images, still so general in Southern Europe, is scouted in England. The importance ascribed to dreams, the barely extinct claims of astrology, and auguries of good or evil luck, and many other well-known products of superstition which are found to exist in every country, have ceased to be believed in by us. This is the natural course of events, just as the Waters of Jealousy and the Urim and Thummin of the Mosaic law had become obsolete in the times of the later Jewish kings. The civilized world has already yielded an enormous amount of honest conviction to the inexorable requirements of solid fact; and it seems to me clear that all belief in the efficacy of prayer, in the sense in which I have been considering it, must be yielded also. The evidence I have been able to collect bears wholly and solely in that direction, and in the face of it the *onus probandi* lies henceforth on the other side.

Nothing that I have said negatives the fact that the mind may be relieved by the utterance of prayer. The impulse to pour out the feelings in sound is not peculiar to Man. Any mother that has lost her young, and wanders about moaning and looking piteously for sympathy, possesses much of that which prompts men to pray in articulate words. There is a yearning of the heart, a craving for help, it knows not where, certainly from no source that it sees. Of a similar kind is the bitter cry of the hare, when the greyhound is almost upon her; she abandons hope through her own efforts, and screams – but to whom? It is a voice convulsively sent out into space, whose utterance is a physical relief. These feelings of distress and of terror are simple, and an inarticulate cry suffices to give vent to them; but the reason why Man is not satisfied by uttering inarticulate cries (though sometimes they are felt to be the most appropriate) is owing to his superior intellectual powers. His memory travels back through interlacing paths, and dwells on various connected incidents; his emotions are complex, and he prays at length.

Neither does anything I have said profess to throw light on the question of how far it is possible for Man to commune in his heart with God. We know that many persons of high intellectual gifts and critical minds look upon it as an axiomatic certainty that they possess this power, although it is impossible for them to establish any satisfactory criterion to distinguish between what may really be borne in upon them from without and what arises from within, but which, through a sham of the imagination, appears to be external. A confident sense of communion with God must necessarily rejoice and strengthen the heart, and divert it from petty cares; and it is equally certain that similar benefits are not excluded from those who on conscientious grounds are sceptical as to the reality of a power of communion. These can dwell on the undoubted fact, that there exists a solidarity between themselves and what surrounds them, through the endless reactions of physical laws, among which the hereditary influences are to be included. They know that they are descended from an endless past, that they have a brotherhood with all that is, and have each his own share of responsibility in the parentage of an endless future. The effort to familiarize the imagination with this great idea has much in common with the effort of communing with a God, and its reaction on the mind of the thinker is in many important respects the same. It may not equally rejoice the heart, but it is quite as powerful in ennobling the resolves, and it is found to give serenity during the trials of life and in the shadow of approaching death.

'ENTHUSIASM'

The changed meaning of the word Enthusiasm is an example of the change of belief in modern times. Its ancient meaning was its literal one, 'God-in-us,' its modern meaning is 'ardent zeal.' I notice that its definition in a recent dictionary is 'a belief or conceit of private revelation; the vain confidence or opinion of a person that he has special divine communications from the Supreme Being, or familiar intercourse with him.' On the other hand, the belief of devout persons that they really commune in their hearts with God, that he put holy ideas and affections into their minds, and inspires them with good resolves, is by no means in their opinion a vain conceit or confidence. To a large number of the ablest class of mankind the idea of an indwelling divine Spirit is so habitual and vivid as to be an axiomatic truth to them. If their views are correct that the germs of a faculty of communing with an unseen world exists in Man, and much more abundantly in some persons than in others, then, considering that this devout persuasion runs in families, as I have fully shown in *Hereditary Genius*, it would follow that those races should be encouraged that are characterized by spiritual-mindedness, and who would be far more worthy occupants of the earth than the generality of ourselves.

It has been to me a real and almost life-long subject of thought, whether or no and to what degree the strong subjective views of the pious are trustworthy. It has been the motive of many of the inquiries in this book, for it seems to me a cardinal one in any question of the improvement of race. Should we keep it before us as an object of endeavour, that future generations may be generally endowed with faculties such as will enable them really to hold as free communion with a Deity as the more spiritually-minded of our race profess to enjoy at the present time? Or is the opinion held by the pious as to the existence of those faculties no more than a vain conceit and confidence, as the dictionary definition just quoted would have it to be?

There is no subject more worthy of reverent but thorough investigation than the objective evidence for or against the existence of inspiration from an unseen world, and none that up to the present time has so tantalized the anxious and honest inquirer with unperformed promises of solution. The arguments scattered or hinted at throughout this book are negative so far as they go, but it must be borne in mind that they would be scattered to the winds by solid objective evidence on the other side, such as could be seriously entertained by scientific men desiring above all things to arrive at truth.

Among the arguments of which I speak, there was evidence that persons in sound health were liable to see visions of an apparently objective character, and to hear voices that seemed external, all of these hallucinations apparently belonging to the same order of phenomena. I also showed that their existing cause could in some instances be traced with more or less certainty; that many of these visions and voices were meaningless or absurd; and that there was not the slightest ground for accrediting the majority of them to any exalted or external source. Similarly, I showed that the fluency of ordinary speakers and writers proceeds in an automatic way, without its being imputed to inspiration, but that when such speakers or writers are exercised upon devout subjects, they are apt to suppose the thoughts that then arise to be inspired, although it would seem to a bystander that all fluency has the same general origin.

I also pointed out that it is among those hysterical or insane persons in whom the sexual organization is disturbed, that the extreme forms of religious rapture chiefly prevail; that the passion of love has many strange metamorphoses, and that life and love in some form, and with its customary illusions, can hardly be separated in a healthy and perfect animal.

An instance of the purely physiological origin of ideas was seen in those twins who are characterized by simultaneity of conceptions; the same notion occurring at the same moment to both, and both responding in nearly the same words and at the same moment to the person who addresses them, so that the twins appear like a double individual.

I have further shown in many ways how little trust can be placed in axiomatic belief. For example, certain natural oddities of mind, such as the perception of number-forms and of colours associated with sounds, always appear to be axiomatic necessities to those who perceive them, and so do many of the sentiments that were instilled in early life. I have also pointed out the necessary untrustworthiness of conscience in some particulars.

Lastly, it appears to be tacitly recognized by all, that the absolute and final court of appeal is not subjective but objective. It is therefore not upon our instinctive convictions or fancies that we should lay most trust, but we should observe the convictions and fancies of others as well as our own, and assign no less trustworthiness to them. Especially should we test the truth of all convictions whenever it is possible to do so, by appeal to such facts as may admit of repetition, for the purpose of verification either by ourselves or by others; experience showing that, in the long run, the supremacy of such facts becomes universally acknowledged. Above all things, we must be content to suspend our belief and maintain the freedom of our mental attitude, wherever there is strong reason for doubt. When there is stormy weather and no secure harbour, the sailors put out to sea; it is not anchorage they then want, but sea-room.

There is nothing in any hesitation that may be felt as to the possibility of receiving help and inspiration from an unseen world, to discredit the practice that is dearly prized by most of us, of withdrawing from the crowd and entering into quiet communion with our hearts, until the agitations of the moment have calmed down, and the distorting mirage of a worldly atmosphere has subsided and the greater objects and more enduring affections of our life have reappeared in their due proportions. We may then take comfort and find support in the sense of our forming part of whatever has existed or will exist, and this need be the motive of no idle reverie but of an active conviction that we possess an influence which may be small but cannot be inappreciable, in defining the as yet undetermined possibilities of an endless future. It may inspire a vigorous resolve to use all the intelligence and perseverance we can command to fulfil our part as members of one great family that strives as a whole towards a fuller and a higher life.

POSSIBILITIES OF THEOCRATIC INTERVENTION

Any attempt to appraise the relative effects of Nature and Nurture may be objected to. It may be said that it is an imperfect and fallacious proceeding to treat the actions of Man as if they were the results of no other influences than may be comprehended under those heads, and that the possibility of theocratic interference must not be overlooked, whether it take place in response to prayer or independently of it. Such an objection may be perfectly valid when the influences at work in any individual case have to be considered, but it happily does not apply to statistical averages for reasons that are quite unconnected with theology, and which I will explain and illustrate. Briefly, it is the very purport and claim of statistics to isolate the effect of specific influences from all others, whether known or unknown, that may act concurrently with them.

Suppose a large number of silk-worms to be tended by a caretaker, and that an observer watched his proceedings as well as he could, but only during the day-time and through a telescope. We will further suppose his observations to show that the worms were of various breeds, and that they were fed in various ways, irrespectively of their breeds, and that the observer desired to discover the relative effects of breed and feeding on the growth of the worms. There can be no doubt as to the principle on which he would work; he would classify his observations so as to compare race with race, and he would reclassify them to compare nurture with nurture. By this well-understood treatment he would isolate the two classes of influence.

Now suppose the caretaker had a custom wholly unknown to the observer, of feeding the worms in various ways during the night-time, how would that affect the statistical conclusions? I answer, only by increasing the amount of individual deviations from the average result, so that, other circumstances remaining the same, the observer would not attain the same constancy in his averages unless the number of observations in his groups was larger than before. Let us consider the ways in which the interference of the caretaker might act.

(1) Suppose he favoured a particular race by giving food to every individual of it during the night-time, then the effect would be that every individual of that race, by virtue of his belonging to the race, would be benefited. The observer who noticed the generally thriving condition of worms of that race would be justified in accepting it as a racial characteristic, for it would be the consequence of the race of the worm.

(2) Suppose the caretaker gave additional food in the night to the particular set whom he had fed the best during the day-time. The observer would rightly ascribe the more or less thriving condition of that set to the peculiarities of their nurture.

(3) Suppose the caretaker acted conversely, feeding those in the night-time whom he had inadequately fed in the day. If the night and day feeding were of equal importance, the observer would find the effects of Nurture to be nil, and rightly so. If they did not balance, he would notice the differential effect.

Thus far we see that the relative total effects of Nature and Nurture would have been rightly appraised. We see at the same time that the effect of any particular kind of Nurture could not be determined, because the whole of the conditions were not under observation.

(4) Suppose the caretaker to feed during the night certain worms that he had marked for the purpose in a manner that wholly escaped the notice of the observer, and that the selection of the worms that were marked had been made on grounds irrespective both of their breed and of the care bestowed on them during the day-time. The result would be that in any large number of worms grouped either according to breed or to the observed dietary, the *proportion* in either group will be the same between those who thrive, and those who otherwise would not have thriven, consequently the *relative* well-being of the two groups will remain unaltered. Favour or disfavour that is bestowed irrespectively of breed and of nurture cannot influence the relative effects of breed and nurture in the long run.

The foregoing arguments cover all composite cases where the influences are mixed, therefore whether there be any unperceived theocratic intervention in favour of particular races, or of individuals irrespectively of race, or partly in one way and partly in another, it cannot under the foregoing suppositions vitiate a statistical comparison between the relative effects of Nature and Nurture, it being understood that 'Nature' refers to all the hereditary gifts and privileges of the race, including constant theocratic intervention in its favour during the period of the observations.

There is, however, a fifth supposition which I feel somewhat ashamed to record. It is that the caretaker, knowing he was watched and not liking it, devised plans for defeating the observer. I fully acknowledge that he could easily succeed in misleading him. The homologue would be a God with the attributes of a Devil, who misled humble and earnest inquirers after truth by malicious artifice. I should not have dared to have alluded to such an ignoble supposition, had not Milton himself put it forward in *Paradise Lost*, Bk. viii, where he makes Raphael tell Adam that God 'did wisely' not to divulge his secrets to be scanned by those who ought rather to admire, and that if they list to conjecture, he has perhaps left the fabric of the heavens to their disputes to 'move his laughter' at their quaint opinions. I think the passage (which was written before Newton's time) must have jarred on the hearts of many readers, and that Milton's supposition of such a character in his God is not likely to be adopted by many persons at the present time. I cannot imagine a more cruel and wicked act, as estimated by the modern instinct of right and wrong, than that which has been so airily suggested by Milton.

We have thus far considered the effects upon statistical conclusions of possible theocratic intervention when given unasked; we have now to consider that which may be accorded in response to petitions. The offering of devout prayer must depend either on the initiative of the Deity or on that of the man. The former condition has just been disposed of under the head of theocratic intervention unasked, the latter can be dealt with in an equally simple manner. The desire to pray, arising independently in the heart of a man, must be due either to his natural character (that is, to his nature), to the external circumstances (all of which I include under the term of his Nurture), or to his free-will. The two first of these are already disposed of, leaving free-will as the only remaining consideration. There are two senses to the word. The popular sense is caprice, or at all events something that acts irrespectively of race and nurture; it therefore falls under the fourth of the conditions already disposed of. Another sense is freedom to follow one's bent, the bent being due either to nature or to circumstances; these cases have also been already considered.

It follows from what has been said that theocratic intervention, whether in response to prayer or given unasked, cannot affect the value of statistical conclusions on the relative total effects of Nature and Nurture, unless Milton's horrible supposition be seriously entertained.

Sir Francis Galton, FRS: Bibliography*

The Telotype: a Printing Electric Telegraph. J. Weale, London (1850).

Extract from a letter of 16 August 1851. *The Times*, 1 January (1852).

'Recent expedition into the interior of South-Western Africa.' *Journal of the Royal Geographical Society*, **22** (1852), pp. 140–63.

Tropical South Africa. John Murray, London (1853). Second edition, under the title *Narrative of an Explorer in Tropical South Africa*, Ward, Lock and Co., London (1889).

'List of astronomical instruments, etc.' In 'Hints to travellers.' *Journal of the Royal Geographical Society*, **24** (1854), pp. 1–13.

'Notes on Modern Geography.' In *Cambridge Essays contributed by Members of the University.* J. W. Parker (ed.), J. W. Parker, London (1885), pp. 79–109.

The Art of Travel; or, Shifts and Contrivances Available in Wild Countries. Murray, London (1855). Second edition (1856). Third edition (1860). Fourth edition (1867). Fifth edition (1872). Sixth edition (1878). Seventh edition (1883). Eighth edition (1893). (David and Charles Reprints (1971), the 1872 edition reprinted under the title *Francis Galton's Art of Travel*, with an introduction by Dorothy Middleton.)

Arts of Campaigning. John Murray, London (1855).

Ways and Means of Campaigning. Privately printed (1855).

Arts of Travelling and Campaigning. T. Brettell, London (1856).

Catalogue of Models Illustrative of the Arts of Camp Life. T. Brettell, London (1858).

'The exploration of arid countries.' *Proceedings of the Royal Geographical Society*, **2** (1858), pp. 60–77.

[Review of *Western Africa*, Hutchinson,] *Proceedings of the Royal Geographical Society*, **2** (1858), pp. 227–9.

'A hand heliostat for the purpose of flashing sun signals, from on board ship or on land, in sunny climates.' *Report of the British Association* (1858), pp. 15–7. Also in *The Engineer*, 15 October (1858), p. 292.

'A description of a hand heliostat.' *Report of the British Association*, (1858), pp. 211–2.

'Sun signals for the use of travellers (hand heliostat).' *Proceedings of the Royal Geographical Society*, **4** (1859), pp. 14–9.

English Weather Data. February 9, 1861, 9 a.m. Privately printed (1861).

'Meteorological charts.' *Philosophical Magazine*, **22** (1861), pp. 34–5.

Circular Letter to Meteorological Observers. Synchronous Weather Charts. Privately printed (1861).

---

* Reprinted from D. W. Forrest, *Francis Galton: The Life and Work of a Victorian Genius* (London: Paul Elek, 1974).

Weather Map of the British Isles for Tuesday, September 3, 1861, 9 a.m. Privately printed (1861).

'Visit to North Spain at the time of the eclipse.' In *Vacation Tourists and Notes of Travel in 1860*. F. Galton (ed.), Macmillan, London (1861), pp. 422–54.

'On a new principle for the protection of riflemen.' *Journal of the Royal United Services Institute*, **4** (1861), pp. 393–6.

'Additional instrumental instructions to Mr Consul Petherick.' *Proceedings of the Royal Geographical Society*, **5** (1861), pp. 96–7.

'Zanzibar.' *The Mission Field*, **6** (1861), pp. 121–30.

'On the "Boussole Burnier", a new French pocket instrument for measuring vertical and horizontal angles.' *British Association Report* (1862), p. 30.

'European weather charts for December 1861', *British Association Report*, (1862), p. 30.

Meteorological Instructions for the Use of Inexperienced Observers Resident Abroad. Meteorological Society, (1862).

'Recent discoveries in Australia.' *The Cornhill Magazine*, **5** (1862), pp. 354–64.

'Report on African explorations.' *Proceedings of the Royal Geographical Society*, **6** (1862), pp. 175–8.

[Preface to] *Vacation Tourists and Notes of Travel in 1861*, F. Galton (ed.), Macmillan, Cambridge and London (1862).

'Explorations in Eastern Africa.' *The Reader*, **1** (1863), p. 19, pp. 42–3. (Signed F. G.).

'The sources of the Nile,' *The Reader*, **1** (1863), p. 615. (Unsigned).

'A development of the theory of cyclones.' *Proceedings of the Royal Society*, **12** (1863), pp. 385–6.

Meteorographica, or Methods of Mapping the Weather. Macmillan, London and Cambridge (1863).

'The climate of Lake Nyanza. Deduced from the observations of Captains Speke and Grant ' *Proceedings of the Royal Geographical Society*, **7** (1863), pp. 225–7.

'The avalanches of the Jungfrau.' *Alpine Journal*, **1** (1863), pp. 184–8.

[Preface to] *Vacation Tourists and Notes of Travel in 1862–3*, F. Galton (ed.), Macmillan, London and Cambridge (1864).

The Knapsack Guide for Travellers in Switzerland. Murray, London (1864).

'First steps towards the domestication of animals.' *British Association Report*, (1864), p. 93–4.

'Captain Speke's new volume.' [Review of *What led to the Discovery of the Source of the Nile*, J. H. Speke,] *The Reader*, **4** (1864), pp. 125–6. (Unsigned).

'Burton on the Nile sources.' [Review of *The Nile Basin*, Richard Burton and James McQueen,] *The Reader*, **4** (1864), p. 728. (Unsigned).

'Grant's Africa.' [Review of *A Walk Across Africa*, James Grant,] *The Reader*, **4** (1864), p. 792. (Unsigned).

'Table for rough triangulation without the usual instruments and without calculation.' *Journal of the Royal Geographical Society*, **34** (1864), pp. 281–4.

'On spectacles for divers, and on the vision of amphibious animals.' *British Association Report*, **35** (1865), pp. 10–11.

[Letter to Major General Sabine on magnetic observations at Tiflis,] *British Association Report*, **35** (1865), pp. 316–7.

[Review of *Frost and Fire, Natural Engines, Tool Marks and Chips*, J. F. Campbell, and of *Ice-caves of France and Switzerland*, C. F. Browne,] *The Edinburgh Review*, **250** (1865), pp. 422–55. (Unsigned).

[Discussion of Nilotic discoveries,] *Proceedings of the Royal Geographical Society*, **9** (1865), pp. 10–11.

'On stereoscopic maps, taken from models of mountainous countries.' *Journal of the Royal Geographical Society*, **35** (1865), pp. 99–104. Summary in *Proceedings of the Royal Geographical Society*, **9** (1865), pp. 104–5.

Hints to Travellers, edited by G. Beck, R. Collinson and F. Galton. Revised edition, Royal Geographical Society, London (1865). Third edition (1871). Fourth edition, edited by Galton alone (1878).

'The first steps towards the domestication of animals.' *Transactions of the Ethnological Society of London*, **3** (1865), pp. 122–38.

'Hereditary talent and character.' *Macmillan's Magazine*, **12** (1865), pp. 157–66, pp. 318–27.

'On an error in the usual method of obtaining meteorological statistics of the ocean.' *British Association Report*, **36** (1866), pp. 16–7. Also in *The Athenaeum*, **2027** (1866), p. 274.

'On the conversion of wind-charts into passage charts.' *British Association Report*, **36** (1866), pp. 17–20. Also in *Philosophical Magazine*, **32** (1866), pp. 345–9.

'Hereditary genius.' [Letter in] *Notes and Queries on China and Japan*. August (1868).

Hereditary Genius. Macmillan, London (1869). Second edition (1892). Third edition, Watts, (1950). Second edition, reprinted with an introduction by C. D. Darlington, Collins, London (1962); and World Publishing Company, New York and Cleveland (1962).

'Description of the pantagraph designed by Mr Galton.' *Minutes of the Meteorological Committee* (1869), p. 9.

'Barometric predictions of weather.' *British Association Report*, **40** (1870), 31–3. Also in *Nature*, **2** (1870), pp. 501–3.

'Mechanical computer of vapour tension.' *Report of the Meteorological Committee* (1871), p. 30.

'Experiments in pangenesis, by breeding from rabbits of a pure variety, into whose circulation blood taken from other varieties had previously been largely transfused.' *Proceedings of the Royal Society*, **19** (1871), pp. 393–410. Also, Taylor and Francis, London (1871).

'Pangenesis.' [Letter in] *Nature*, **4** (1871), pp. 5–6.

'Gregariousness in cattle and in men.' *Macmillan's Magazine*, **23** (1871), pp. 353–57.

'Statistical inquiries into the efficacy of prayer.' *Fortnightly Review*, **12** (1872), pp. 125–35. Also in *The Prayer Gauge Debate*, J. O. Means (ed.), Congregational Publishing Co., Boston (1876).

'The efficacy of prayer.' [Letter in] *Spectator*, 24 August 1872, p. 1073.

'On blood-relationship.' *Proceedings of the Royal Society*, **20** (1872), pp. 394–402. Also in *Nature*, **6** (1872), pp. 173–6.

[Opening address by the president to the Geographical Section.] *British*

Association Report, **62** (1872), pp. 198–203. Also, in *Nature*, **6** (1872), pp. 343–5. Also William Clowes and Sons, Brighton (1872).

'Memorandum on the construction of "isodic charts", by which the average length of a day's sail by any particular class of vessels, in any direction, at any place, and in any season, may be readily found, and the average duration of passages along suggested tracks determined and compared.' *Minutes of the Meteorological Council*, 2 December (1872).

[Letter in] *Daily News*, 7 September (1872).

'Lieutenant Dawson and his barometers.' [Letter in] *Daily News*, 18 September (1872).

'Barometer tubes.' [Letter in] *Daily News*, 20 September (1872).

'On the employment of meteorological statistics in determining the best course for a ship whose sailing qualities are known.' *Proceedings of the Royal Society*, **21** (1873), pp. 263–74.

Hereditary improvement.' *Fraser's Magazine*, **7** (1873), pp. 116–30.

'On the extinction of surnames.' *Educational Times* (1873).

'On a proposed stable state.' *Nature*, **9** (1873), pp. 342–3.

'The relative supplies from town and country families to the population of future generations.' *Journal of the Statistical Society*, **36** (1873), pp. 19–26.

'On the causes which operate to create scientific men.' *Fortnightly Review*, **13** (1873), pp. 345–51.

'Africa for the Chinese.' [Letter in] *The Times*, 6 June (1873).

English Men of Science: their Nature and Nurture. Macmillan, London (1874), Second edition, with an introduction by Ruth Schwartz Cohen, Frank Cass, London (1970).

'On men of science, their nature and their nurture.' *Proceedings of the Royal Institution*, **7** (1874), pp. 227–36. Also in *Nature*, **9** (1874), pp. 344–5.

'On a proposed statistical scale.' [Letter in] *Nature*, **9** (1874), pp. 342–3.

'Nuts and men.' [Letter in] *Spectator*, 30 May (1874), p. 689.

'Proposal to apply for anthropological statistics from schools.' *Journal of the Anthropological Institute*, **3** (1874), pp. 308–11.

'Notes on the Marlborough School statistics.' *Journal of the Anthropological Institute*, **4** (1874), pp. 130–5.

'On the excess of females in the West Indian Islands, from documents communicated to the Anthropological Institute of the Colonial Office.' *Journal of the Anthropological Institute*, **4** (1874), pp. 136–7.

[Prefatory remarks to 'On the probability of the extinction of families', H. W. Watson,] *Journal of the Anthropological Institute*, **4** (1874), pp. 138–9.

[Review of *Heredity; a Psychological Study of its Phenomena, Laws, Causes, and Consequences*, Th. Ribot,] *The Academy*, 30 January (1875).

'Statistics by intercomparison with remarks on the Law of Frequency of Error.' *Philosophical Magazine*, **49** (1875), pp. 33–46.

[Discussion of 'Ultra-Centenarian Longevity', G. D. Gibb,] *Journal of the Anthropological Institute*, **5** (1875), pp. 98–9.

'The history of twins, as a criterion of the relative powers of nature and nurture.' *Fraser's Magazine*, **12** (1875), pp. 566–76. Revised version reprinted in *Journal of the Anthropological Institute*, **5** (1875), pp. 391–406. Summary in *Nature*, **13** (1875), p. 59.

'Short notes on heredity etc., in twins.' *Journal of the Anthropological*

Institute, **5** (1875), pp. 324–9. Extracts reprinted under the title 'Twins and fertility'. *The Live Stock Journal and Fanciers' Gazette*, **3** (1876), p. 148.

'A theory of heredity.' *Contemporary Review*, **27** (1875), pp. 80–95. Also, under the title 'Théorie de l'hérédité'. *La Revue Scientifique*, **10** (1876), pp. 198–205. Revised version reprinted in *Journal of the Anthropological Institute*, **5** (1875), pp. 329–48. Summary in *Nature*, **13** (1875), p. 59.

'On the height and weight of boys aged 14, in town and country public schools.' *Journal of the Anthropological Institute*, **6** (1876), pp. 174–80.

[Report on measurement in] *The Effects of Cross and Self-Fertilization in the Animal Kingdom*, C. Darwin, Murray, London (1876), pp. 16–8.

'Whistles for determining the upper limits of audible sound in different persons.' In *Physics and Mechanics, South Kensington Museum Conference*. Chapman and Hall, London (1876).

'Typical laws of heredity.' *Proceedings of the Royal Institution*, **8** (1877), 282–301. Also in *Nature*, **15** (1877), pp. 492–5, pp. 512–4, pp. 532–3; and under the title 'Les lois typiques de l'hérédité', *La Revue Scientifique*, **13** (1877), pp. 385–93.

'Considerations adverse to the maintenance of Section F.' *Journal of the Statistical Society*, **40** (1877), pp. 468–73.

'Address to the Department of Anthropology, Section H.' *British Association Report*, (1877), pp. 94–100. Summary in *Nature*, **16** (1877), pp. 344–7. Also, under the title *Address to the Anthropological Department of the British Association*. W. Clowes and Sons, London (1877); and under the title 'La psycho-physique', *La Revue Scientifique*, **13** (1877), pp. 494–8.

'Bicycle speedometer.' *Field* (1877).

'Description of the process of verifying thermometers at the Kew Observatory.' *Proceedings of the Royal Society*, **26** (1877), pp. 84–9. Also in *Philosophical Magazine*, **4** (1877), pp. 226–31.

'On means of combining various data in maps and diagrams,' in *Chemistry, Biology, etc., South Kensington Museum Conference*. Chapman and Hall, London (1878), pp. 312–5.

'Letters of Henry Stanley from Equatorial Africa to the *Daily Telegraph*.' *Edinburgh Review*, January (1878), pp. 166–91. (Unsigned).

'Composite portraits made by combining those of many different persons into a single figure,' *Journal of the Anthropological Institute*, **8** (1878), pp. 132–48. Also, under the title 'Les portraits composites', *La Revue Scientifique*, **15** (1878), pp. 33–8; and *Nature*, **18** (1878), pp. 97–100; and Harrison and Sons, London (1878).

'On the advancement of geographical teaching.' *Nature*, **18** (1878), p. 337.

'Psychometric facts.' *Nineteenth Century*, March (1879), pp. 425–33.

'Psychometric experiments.' *Brain*, **2** (1879), pp. 149–62. Also William Clowes & Sons, London (1879).

'Generic images.' *Proceedings of the Royal Institution*, **9** (1879), 161–70. Also, William Clowes and Sons, London (1879). With additions, under the title 'Les images génériques', *La Revue Scientifique*, **17** (1879), pp. 221–5.

'Generic images.' *Nineteenth Century*, **6** (1879), pp. 157–69.

'The average flush of excitement.' [Letter in] *Nature*, **20** (1879), p. 121.

'The geometric mean, in vital and social statistics.' *Proceedings of the Royal Society*, **29** (1879), pp. 365–7.

'Statistics of mental imagery.' *Mind*, **5** (1880), pp. 301–18.

'Mental imagery,' *Fortnightly Review*, **28** (1880), pp. 312–24.
'Visualised numerals.' *Nature*, **21** (1880), pp. 252–6, 494–5.
'Visualised numerals.' [Letter in] *Nature*, **21** (1880), p. 323.
'Visualised numerals.' *Journal of the Anthropological Institute*, **10** (1880), pp. 85–102. Also Harrison and Sons, London (1880).
'The opportunities of science masters at schools.' [Letter in] *Nature*, **22** (1880), pp. 9–10.
'On determining the heights and distances of clouds by their reflexions in a low pond of water, and in a mercurial horizon.' *British Association Report*, **50** (1880), pp. 459–61.
'On a pocket registrator for anthropological purposes.' *British Association Report*, **50** (1880), p. 625. Summary in *Nature*, **22** (1880), p. 478.
'The visions of sane persons.' *Fortnightly Review*, **29** (1881), pp. 729–40. Reprinted with slight variations in *Proceedings of the Royal Institution*, **9** (1881), pp. 644–55.
'Composite portraiture.' *Photographic Journal*, **5** (1881), pp. 140–6.
'Composite portraiture.' *Photographic News*, **25** (1881), pp. 316–7, pp. 332–3.
'On the application of composite portraiture to anthropological purposes.' *British Association Report*, **51** (1881), p. 3.
[Tables and discussion of range in height, weight and strength] in Report of the Anthropometric Committee, *British Association Report*, **51** (1881), pp. 225–72.
'On the construction of isochronic passage charts.' *British Association Report*, **51** (1881), pp. 740–1. Also, *Proceedings of the Royal Geographical Society*, **3** (1881), pp. 657–8. (A chart appears as frontispiece to *Hints to Travellers*, fifth edition, 1883.)
[Discussion on 'On the laws affecting the relations between civilized and savage life as bearing on the dealings of colonists with aborigines', H. Bartle Frere] *Journal of the Anthropological Institute*, **11** (1881), pp. 352–3.
'Burials in the Abbey.' [Letter in] *Pall Mall Gazette*, 27 December (1881).
'Photographic chronicles from childhood to age.' *Fortnightly Review*, **181** (1882), pp. 26–31.
With F. A. Mahomed. 'An inquiry into the physiognomy of phthisis by the method of "composite portraiture".' *Guy's Hospital Reports*, 25 February (1882).
'The late Mr Darwin: a suggestion.' [Letter in] *Pall Mall Gazette*, 27 April (1882).
'Conventional representation of the horse in motion.' *Nature*, **26** (1882), pp. 228–9.
'A rapid-view instrument for momentary attitudes.' *Nature*, **26** (1882), pp. 249–51.
[Discussion on 'Analysis of relationships of consanguinity and affinity', A. McFarlane,] *Journal of the Anthropological Institute*, **12** (1882), pp. 61–2.
'The anthropometric laboratory.' *Fortnightly Review*, **183** (1882), pp. 332–8.
Inquiries into Human Faculty and Its Development. Macmillan, London (1883). Second edition, Macmillan (1892). Third edition, Dent (1907). Fourth edition, Eugenics Society (1951).
'Hydrogen whistles.' *Nature*, ·**27** (1883), pp. 491–2. Also, corrected, in *Nature*, **28** (1883), p. 54.

'Method of determining the distance and height of clouds and the direction and rate of their motions parallel to the earth's surface.' *Meteorological Council*, April (1883).

'The American trotting-horse.' *Nature*, **28** (1883), p. 29.

[Reply to Romanes' review of *Human Faculty and Its Development*,] *Nature*, **28** (1883), pp. 97–8.

'Arithmetic notation of kinship.' [Letter in] *Nature*, **28** (1883), p. 435.

Outfit for an anthropometric laboratory. Privately printed (1883).

'On apparatus for testing the delicacy of the muscular and other senses in different persons.' *Journal of the Anthropological Institute*, **12** (1883), pp. 469–77.

[Obituary: William Spottiswoode.] *Proceedings of the Royal Geographical Society*, **6** (1883), pp. 489–91.

'Final Report of the Anthropometric Committee.' *British Association Report* (1883), pp. 253–306.

'Report of the Local Scientific Societies Committee.' *British Association Report* (1883), pp. 318–45. Partly reprinted in *Nature*, **28** (1883), pp. 135–6.

'Family records.' [Letter in] *The Times*, 9 January (1884). Also in *Nature*, **30** (1884), p. 82.

'The weights of British noblemen during the last three generations.' *Nature*, **29** (1884), pp. 266–8.

[Discussion on 'On the races of the Congo and the Portuguese colonies in Western Africa', H. H. Johnston,] *Journal of the Anthropological Institute*, **13** (1884), pp. 478–9.

Anthropometric Laboratory. William Clowes and Sons, London (1884).

'On the Anthropometric Laboratory at the late International Health Exhibition.' *Journal of the Anthropological Institute*, **14** (1884), pp. 205–18. Also Harrison and Sons, London (1885).

'Some results of the Anthropometric Laboratory.' *Journal of the Anthropological Institute*, **14** (1884), pp. 275–87.

Record of Family Faculties. Macmillan, London (1884).

Life-History Album. Macmillan, London (1884). First edition edited by F. Galton. Second edition, rearranged by F. Galton (1902).

'Free-will—observations and inferences.' *Mind*, **9** (1884), pp. 406–13.

'Measurement of character.' *Fortnightly Review*, **36** (1884), pp. 179–85.

'The identiscope.' *Nature*, **30** (1884), pp. 637–8.

'The cost of anthropometric measurements.' [Letter in] *Nature*, **31** (1884), p. 150.

'Anthropometric percentiles.' *Nature*, **31** (1885), pp. 223–5.

'Hereditary deafness.' [Review of 'Upon the formation of a deaf variety of the human race', Alexander Bell,] *Nature*, **31** (1885), pp. 269–70.

'Photographic composites.' *Photographic News*, **29** (1885), pp. 234–45.

'A common error in statistics.' *Jubilee Volume of the Statistical Society*. Edward Stanford, London (1885), p. 261.

'The application of a graphic method to fallible measures.' *Jubilee Volume of the Statistical Society*, Edward Stanford, London (1885), pp. 262–5.

'The measure of fidget.' *Nature*, **32** (1885), pp. 174–5.

[Presidential address, Section H, Anthropology,] *British Association Report*,

55 (1885), pp. 1206–14. Also, under the title *Address to the Section of Anthropology of the British Association*. Spottiswoode, London (1885); and, *Nature*, **32** (1885), pp. 507–10.

'Regression towards mediocrity in hereditary stature.' *Journal of the Anthropological Institute*, **15** (1885), pp. 246–63.

[Opening remarks by the President, and discussion on 'Experiments in testing the character of school children,' Sophie Bryant,] *Journal of the Anthropological Institute*, **15** (1885), pp. 336–8, p. 350.

'Exhibition of composite photographs of skulls by Francis Galton.' *Journal of the Anthropological Institute*, **15** (1885), pp. 390–1.

'Conference of delegates of corresponding societies of the British Association held at Aberdeen.' *Nature*, **33** (1885), pp. 81–3.

'Hereditary Stature.' [Extracts from Presidential address to the Anthropological Institute.] *Nature*, **33** (1885), pp. 295–8.

'Hereditary Stature.' [Letter in] *Nature*, **33** (1885), p. 317.

[President's address,] *Journal of the Anthropological Institute*, **15** (1886), pp. 489–99.

'On recent designs for anthropometric instruments.' *Journal of the Anthropological Institute*, **16** (1886), pp. 2–8.

'Family likeness in stature.' *Proceedings of the Royal Society*, **40** (1886), pp. 42–63.

[On *American Family Peculiarities in the Eighteenth Century*, J. Boucher,] *Journal of the Anthropological Institute*, **16** (1886), pp. 98–9.

'Notes on permanent colour types in mosaic.' *Journal of the Anthropological Institute*, **16** (1886), pp. 145–7. Also, Harrison and Sons, London (1886).

'Family-likeness in eye-colour.' *Proceedings of the Royal Society*, **40** (1886), pp. 402–16. Also, summary in *Nature*, **34** (1886), p. 137.

[Opening remarks by the President,] *Journal of the Anthropological Institute*, **16** (1886), pp. 175–7, pp. 189–90.

'The origin of varieties.' *Nature*, **34** (1886), pp. 395–6.

'Supplementary notes on "Prehension in idiots".' *Mind*, **12** (1886), pp. 79–82.

[Galton's speech at the Royal Society dinner after receiving the Gold Medal of the Society,] *The Times*, 1 December (1886).

'Chance and its bearing in heredity.' *Birmingham Daily Post*, 7 December (1886).

Pedigree Moths. On a Proposed Series of Experiments in Breeding Moths. Privately printed (1887).

[Address delivered at the anniversary meeting of the Anthropological Institute of Great Britain and Ireland,] *Journal of the Anthropological Institute*, **16** (1887), pp. 387–402. Also Harrison and Sons, London (1887).

'Thoughts without words.' [Letters in] *Nature*, **36** (1887), pp. 28–9, pp. 100–1.

'North American pictographs.' *Nature*, **36** (1887), pp. 155–7.

List of Anthropometric Apparatus. Cambridge Scientific Instrument Company, Cambridge (1887).

'Good and bad temper in English families.' *Fortnightly Review*, **42** (1887), pp. 21–30.

'Photography and silhouettes.' [Letters in] *Photographic News*, **31** (1887), pp. 429–30, p. 462.

'Pedigree moth-breeding as a means of verifying certain important constants in the general theory of heredity.' *Transactions of the Entomological Society*, Part I (1887), pp. 19–34.

[Discussion on 'On an Ancient British Settlement excavated near Rushmore, Salisbury', Pitt Rivers,] *Journal of the Anthropological Institute*, **17** (1887), 199–200.

'The proposed Imperial Institute: geography and anthropology.' [Letter in] *The Times*, 6 October (1887).

[Address delivered at the anniversary meeting of the Anthropological Institute of Great Britain and Ireland,] *Journal of the Anthropological Institute*, **17** (1887), 346–54. Also Harrison and Sons, London (1888).

'Note on Australian marriage systems.' *Journal of the Anthropological Institute*, **18** (1887), 70–2. Also Harrison and Sons, London (1888).

'Composite portraiture. A communication from Francis Galton.' *Photographic News*, **32** (1888), 257.

'On head growth in students at the University of Cambridge.' *Journal of the Anthropological Institute*, **18** (1888), 155–6. Also in *Nature*, **38** (1888), 14–5.

'Remarks on replies by teachers to questions respecting mental fatigue.' *Journal of the Anthropological Institute*, **18** (1888), 157–68. Also Harrison and Sons, London (1888).

'Personal identification and description.' *Nature*, **38** (1888), 173–7, 201–2. Also in *Journal of the Anthropological Institute*, **18** (1888), 177–91; and revised version in *Proceedings of the Royal Institution*, **12** (1888), 346–60.

[Discussion on 'On a method of investigating the development of institutions applied to laws of marriage and descent', E. Tylor,] *Journal of the Anthropological Institute*, **18** (1888), p. 270.

[Remarks on the 'Exhibition of an ancient Peruvian gold breastplate',] *Journal of the Anthropological Institute*, **18** (1888), p. 274.

'Co-relations and their measurement, chiefly from anthropometric data.' *Proceedings of the Royal Society*, **45** (1888), pp. 135–45. Also in *Nature*, **39** (1889), p. 238.

'Table of observations.' *Journal of the Anthropological Institute*, **18** (1889), pp. 420–30. Also Harrison and Sons, London (1889).

Natural Inheritance. Macmillan, London (1889).

[Address delivered at the annual meeting of the Anthropological Institute of Great Britain and Ireland,] *Journal of the Anthropological Institute*, **18** (1889), pp. 401–19. Also in *Nature*, **39** (1889), pp. 296–7; and Harrison and Sons, London (1889).

'The sacrifice of education to examination.' *Nineteenth Century*, **25** (1889), pp. 303–8.

'Exhibition of instruments (1) for testing the perception of differences of tint, and (2) for determining reaction-time.' *Journal of the Anthropological Institute*, **19** (1889), pp. 27–9. Also Harrison and Sons, London (1889).

'Head growth in students at the University of Cambridge.' [Letter in] *Nature*, **40** (1889), p. 318.

'On the advisability of assigning marks for bodily efficiency in the examination of candidates for public services.' *British Association Report,* **59** (1889), pp. 471–3.

'On the principle and methods of assigning marks for bodily efficiency.' *British Association Report,* **59** (1889), pp. 474–8. Also in *Nature,* **40** (1889), pp. 631–2, pp. 649–52.

'Feasible experiments on the possibility of transmitting acquired habits by means of inheritance.' *British Association Report,* **59** (1889), pp. 620–1. Also in *Nature,* **40** (1889), p. 610.

'An instrument for measuring reaction time.' *British Association Report,* **59** (1889), pp. 784–5.

'Head measures at Cambridge.' [Letter in] *Nature,* **40** (1889), p. 643.

'Why do we measure mankind?' *Lippincott's Monthly Magazine,* **45** (1890), pp. 236–41.

'Cambridge anthropometry.' *Nature,* **41** (1890), p. 454.

Tests and Certificates of the Kew Observatory. Kew Committee of the Royal Society (1890).

'Dice for statistical experiments.' *Nature,* **42** (1890), pp. 13–4.

[Review of *The Criminal,* Havelock Ellis,] *Nature,* **42** (1890), pp. 75–6.

'A new instrument for measuring the rate of movement of the various limbs.' *Journal of the Anthropological Institute,* **20** (1890), 200–4. Also, summary in *Nature,* **42** (1890), p. 143.

'Physical tests in competitive examinations.' *Journal of the Society of Arts,* **39** (1890), pp. 19–27.

[Remarks following 'Mental tests and measurements', J. McK. Cattell,] *Mind,* **15** (1890), pp. 380–1.

Anthropometric Laboratory. Notes and Memoirs No. 1. Richard Clay, London (1890).

[Obituary: The Reverend G. Butler, D.D.,] *Proceedings of the Royal Geographical Society,* **12** (1890), pp. 236–7.

[Discussion on 'An apparent paradox in mental evolution', Lady Welby,] *Journal of the Anthropological Institute,* **20** (1890), pp. 304–23.

'The patterns in thumb and finger marks.' *Proceedings of the Royal Society,* **48** (1890), pp. 455–7. Also in *Journal of the Anthropological Institute,* **20** (1890), pp. 360–1; and *Nature,* **43** (1890), pp. 117–8.

[Obituary: Miss North,] *Journal of the Anthropological Institute,* **20** (1891), p. 302.

'Methods of indexing finger-marks.' *Proceedings of the Royal Society,* **49** (1891), pp. 540–48. Also in *Nature,* **44** (1891), p. 141.

'Meteorological phenomena.' [Letter in] *Nature,* **44** (1891), p. 294.

'Identification by finger tips.' *Nineteenth Century,* **30** (1891), pp. 303–11.

'Retrospect of work done at my anthropometric laboratory at South Kensington.' *Journal of the Anthropological Institute,* **21** (1891), pp. 32–5. Also Harrison and Sons, London (1891).

'The patterns in thumb and finger marks.' *Philosophical Transactions, B,* **182** (1891), pp. 1–23.

[Reminiscences of Henry Walter Bates, F.R.S.,] *Proceedings of the Royal Geographical Society,* **14** (1892), pp. 255–7.

[Presidential address to the Division of Demography,] *Transactions of the Seventh International Congress of Hygiene and Demography*, **10** (1892), pp. 7–12.

'Finger prints and their registration as a means of personal identification.' *Transactions of the Seventh International Congress of Hygiene and Demography*, **10** (1892), pp. 301–3.

Finger Prints. Macmillan, London (1892). Reprinted, Da Capo Press, New York (1965), with an introduction by H. Cummins.

[Sections on 'Causes that limit population', 'astronomy', 'communications', 'statistics', 'population',] in *Notes and Queries on Anthropology*, Pitt-Rivers (ed.), second edition. British Association for the Advancement of Science (1892), pp. 204, 208, 221, 226, 229.

'The just-perceptible difference.' *Proceedings of the Royal Institution*, **14** (1893), pp. 13–26. Also, extracts published under the title 'Measure of the imagination.' *Nature*, **47** (1893), pp. 319–21; and 'Optical continuity.' *Nature*, **47** (1893), pp. 342–5.

'Enlarged finger prints.' *Photographic Work*, 10 February (1893).

'Identification.' [Letter in] *The Times*, 7 July (1893).

'Identification.' [Letter in] *Nature*, **48** (1893), p. 222.

'Recent introduction into the Indian Army of the method of finger-prints for the identification of recruits.' *British Association Report* (1893), p. 902. Also under the title 'Finger prints in the Indian Army.' *Nature*, **48** (1893), p. 595.

Decipherment of Blurred Finger Prints. Macmillan, London (1893).

'Payments through telegram by P.O. Savings books.' [Letter in] *The Times*, 27 December (1893).

'Arithmetic by smell.' *Psychological Review*, **1** (1894), pp. 61–2.

'A plausible paradox in chances.' *Nature*, **49** (1894), pp. 365–6.

'Results derived from the natality tables of Korosi by employing the method of contours or isogens.' *Proceedings of the Royal Society*, **55** (1894), pp. 18–23. Also in *Nature*, **49** (1894), p. 570.

'The relative sensitivity of men and women at the nape of the neck, by Weber's test,' *Nature*, **50** (1894), pp. 40–2.

'Discontinuity in evolution.' *Mind*, **3** (1894), pp. 362–72.

Physical Index to 100 Persons Based on their Measures and Finger Prints. Privately printed (1894).

'The part of religion in human evolution.' *National Review*, **23** (1894), pp. 755–63.

'Acquired characters.' [Letter in] *Nature*, **51** (1894), p. 56.

'Psychology of mental arithmeticians and blindfold chess-players.' [Review of *Psychologie des Grands Calculateurs et Joueurs d'Echecs*, Alfred Binet,] *Nature*, **51** (1894), pp. 73–4.

'A new step in statistical science.' [Letter in] *Nature*, **51** (1895), p. 319.

'Questions bearing on specific stability.' *Transactions of the Entomological Society of London* (1895), pp. 155–7. Also in *Nature*, **51** (1895), pp. 570–1.

Finger Print Directories. Macmillan, London (1895).

'Terms of imprisonment.' *Nature*, **52** (1895), pp. 174–6.

'Personality.' [Review of *The Diseases of Personality*, Th. Ribot,] *Nature*, **52** (1895), pp. 517–8.

'The wonders of a finger print.' *Sketch*, 20 November (1895).

'Prints of scars.' [Letter in] *Nature*, **53** (1896), p. 295.

[Obituary notices of Fellows deceased: Dr John Rae,] *Proceedings of the Royal Society*, **60** (1896), pp. 5–7.

'Les empreintes digitales', *Comptes-rendus du IVe Congres International d'Anthropologie Criminelle*. Sessions de Genève (1896), pp. 35–8.

'Three generations of lunatic cats.' *Spectator*, 11 April (1896).

'A curious idiosyncrasy.' *Nature*, **54** (1896), p. 76.

'Intelligible signals between neighbouring stars.' *Fortnightly Review*, **60** (1896), pp. 657–64.

'The Bertillon system of identification'. [Review of *Signaletic Instructions*, Alphonse Bertillon,] *Nature*, **54** (1896), pp. 569–70.

Private circular of Committee for Measurement of Plants and Animals. Royal Society, 30 November (1896).

'Rate of racial change that accompanies different degrees of severity in selection.' *Nature*, **55** (1897), pp. 605–6.

'Note to the memoir by Professor Karl Pearson, F.R.S., on spurious correlation.' *Proceedings of the Royal Society*, **60** (1897), pp. 498–502.

'Retrograde selection.' *Gardiners' Chronicle*, 15 May (1897).

'The average contribution of each several ancestor to the total heritage of the offspring.' *Proceedings of the Royal Society*, **61** (1897), pp. 401–13. Also, summary in 'A new law of heredity.' *Nature*, **56** (1897), pp. 235–7.

'Hereditary colour in horses.' *Nature*, **56** (1897), pp. 598–9.

'Relation between individual and racial variability.' [Unsigned review of 'A measure of variability and the relation of individual variation to specific differences', E. T. Brewster,] *Nature*, **57** (1897), pp. 16–7.

'The late Dr. Haughton.' [Letter in] *Nature*, **57** (1897), p. 79.

'An examination into the registered speeds of American trotting horses, with remarks on their value as hereditary data.' *Proceedings of the Royal Society*, **62** (1897), pp. 310–5. Also in *Nature*, **58** (1898), pp. 333–4.

'Photographic measurement of horses and other animals.' *Nature*, **57** (1898), pp. 230–2.

'A diagram of heredity.' [Letter in] *Nature*, **57** (1898), p. 292.

'Temporary flooring in Westminster Abbey for ceremonial procession.' [Letter in] *The Times*, 25 May (1898).

'Evolution of the moral instinct.' [Review of *The Origin and Growth of the Moral Instinct*, Alexander Sutherland,] *Nature*, **58** (1898), pp. 241–2. (Signed F. G.).

'The distribution of prepotency.' [Letter in] *Nature*, **58** (1898), pp. 246–7.

'Photographic records of pedigree stock.' [Letter in] *Live Stock Journal*, 30 September (1898).

'Corporal punishment.' [Letter in] *The Times*, 4 October (1898). (Signed F. G.).

'Photographic records of pedigree stock.' *British Association Report*, (1898), pp. 597–603. Summary in *Nature*, **58** (1898), p. 584.

'The photography of the premium horses.' *Appendix G, 7th Report of the Royal Commission on Horse Breeding*, (1899), pp. 12–3.

'A measure of the intensity of hereditary transmission.' [Letter in] *Nature*, **60** (1899), p. 29.

'Strawberry cure for gout.' [Letter in] *Nature*, **60** (1899), p. 125.
'Pedigree stock records.' *British Association Report*, **69** (1899), pp. 424–9.
'The median estimate.' *British Association Report*, **69** (1899), pp. 638–40. Summary in *Nature*, **60** (1899), p. 584.
'Finger prints of young children.' *British Association Report*, **69** (1899), pp. 868–9.
'A geometric determination of the median value of a system of normal variants from two of its centiles.' *Nature*, **61** (1899), pp. 102–4.
[Discussion on 'The metric system of identification of criminals as used in Great Britain and Ireland', J. G. Carson,] *Journal of the Anthropological Institute*, **30** (1900), pp. 195–6.
[Introduction to] *William Cotton Oswell: The Story of His Life*, W. E. Oswell, Heinemann, London (1900).
'Souvenirs d'Egypte.' *Bulletin de la Société Khédiviale de Géographie*, Ve Serie, **7** (1900), pp. 375–80.
'Analytical portraiture.' [Letter in] *Nature*, **62** (1900), p. 320.
'Analytical photography.' *Photographic Journal*, **25** (1900), pp. 135–8.
'Identification offices in India and Egypt.' *Nineteenth Century*, **48** (1900), pp. 118–26.
'Biometry.' *Biometrika*, **1** (1901), pp. 7–10.
'The possible improvement of the human breed under the existing conditions of law and sentiment.' *Nature*, **64** (1901), pp. 659–65. Also in *Report of the Smithsonian Institution* (1901), pp. 523–38.
'On the probability that the son of a very highly-gifted father will be no less gifted.' *Nature*, **65** (1901), p. 79.
'The most suitable proportion between the values of first and second prizes.' *Biometrika*, **1** (1902), pp. 380–90.
[Letter in] *Truth*, **52** (1902), p. 786.
'Finger print evidence.' *Nature*, **66** (1902), p. 606.
'Sir Edward Fry and natural selection.' [Letter in] *Nature*, **67** (1903), p. 343.
'Pedigrees, based on fraternal unities.' *Nature*, **67** (1903), pp. 586–7.
'Our national physique—prospects of the British race—are we degenerating?' Article in *Daily Chronicle*, 29 July (1903).
'Nomenclature and tables of kinship.' *Nature*, **69** (1904), pp. 294–5.
'African memorial.' [Letter in] *The Times*, 25 May (1904).
'Eugenics. Its definition, scope and aims.' *Nature*, **70** (1904), p. 82. Also in *Sociological Papers*, **1** (1905), pp. 45–50, pp. 78–9.
'Distribution of successes and of natural ability among the kinsfolk of Fellows of the Royal Society.' *Nature*, **70** (1904). pp. 354–6.
'Average number of kinsfolk in each degree.' *Nature*, **70** (1904), p. 529, p. 626, and **71** (1905), p. 248.
'On the character and ancestry of Lord Northbrook.' [Letter in] *The Times*, 17 November (1904).
'A eugenics investigation: index to achievements of near kinsfolk of some of the Fellows of the Royal Society.' *Sociological Papers*, **1** (1905), pp. 85–9. Also R. Clay, London (1905).
'Studies in national eugenics.' *Nature*, **71** (1905), pp. 401–2.
'Number of strokes of the brush in a picture.' [Letter in] *Nature*, **72** (1905), p. 198.

[Review of] *Guide to Finger Print Identification*, Henry Faulds, *Nature*, **72** (1905), Supplement iv–v.

'Nomenclature of kinship—its extension.' [Letter in] *Nature*, **73** (1905), pp. 150–1.

'Eugenics: I. Restrictions in marriage; II. Studies in national eugenics.' *Sociological Papers*, **2** (1906), pp. 3–13, pp. 14–17, pp. 49–51.

Noteworthy Families. With E. Schuster. John Murray, London (1906).

'Request for prints of photographic portraits.' [Letter in] *Nature*, **73** (1906), p. 534.

'Anthropometry at schools.' *Journal of preventive Medicine*, **14** (1906), pp. 93–8.

'Measurement of resemblance.' [Letter in] *Nature*, **74** (1906), pp. 562–3.

'Cutting a round cake on scientific principles.' [Letter in] *Nature*, **75** (1906), p. 173.

'One vote, one value.' [Letter in] *Nature*, **75** (1907), p. 414.

'Vox populi.' *Nature*, **75** (1907), pp. 450–1.

'The ballot box.' [Letter in] *Nature*, **75** (1907), pp. 509–10.

Probability, the Foundation of Eugenics. Henry Froude, London (1907).

'Grades and deviates.' *Biometrika*, **5** (1907), pp. 400–4.

'Classification of portraits', *Nature*, **76** (1907), pp. 617–8.

'Suggestions for improving the literary style of scientific memoirs.' *Transactions of the Royal Society of Literature*, **28** (1908), Part II, pp. 1–8.

'Address on eugenics.' *Westminster Gazette*, 26 June (1908).

Memories of My Life. Methuen, London (1908).

'Local associations for promoting eugenics.' *Nature*, **78** (1908), pp. 645–7.

'Identification by finger prints.' [Letter in] *The Times*, 13 January (1909).

'Sequestrated church property.' [Letter in] *Nature*, **79** (1909), p. 308.

[Foreword.] *Eugenics Review*, **1** (1909), pp. 1–2.

[Preface to] *Primer of Statistics*, W. P. and Ethel M. Elderton, Black, London (1909).

'Deterioration of the British Race.' [Letter in] *The Times*, 18 June (1909).

'Segregation (of the feeble-minded)' in *The Problem of the Feeble Minded*. P. S. King, London (1909).

Essays in Eugenics. Eugenics Education Society, London (1909).

'Eugenic qualities of primary importance.' *Eugenics Review*, **1** (1910), pp. 74–6.

'Note on the effects of small and persistent influences.' *Eugenics Review*, **1** (1910), pp. 148–9.

'Numeralised profiles for classification and recognition.' *Nature*, **83** (1910), pp. 127–30.

'Heredity and tradition.' [Letter in] *The Times*, 31 May (1910).

'Alcoholism and offspring.' [Letter in] *The Times*, 3 June (1910).

'Eugenics and the Jew.' [Letter in] *Jewish Chronicle*, 30 July (1910).

'The Eugenics Laboratory and the Eugenics Education Society.' [Letter in] *The Times*, 3 November (1910).

List of Presidents

Eugenics Education Society, 1907–1926

Sir Francis Galton, FRS (1822–1911), Honorary President, 1908–11
Sir James Crighton-Browne, MD, FRS (1840–1938), 1908–9
Mr Montague Crackanthorpe (1832–1913), 1909–11
Major Leonard Darwin (1850–1943), 1911–26

Eugenics Society, 1926–1989

Major Leonard Darwin (1850–1943), 1926–9
Sir Bernard Mallet (1850–1932), 1929–32
Sir Humphry Rolleston, MD (1862–1944), 1933–5
Lord Horder, MD (1871–1955), 1935–48
Sir Alexander Carr-Saunders (1886–1966), 1949–53
Sir Charles Darwin, FRS (1887–1962), 1954–9
Sir Julian Huxley, FRS (1887–1975), 1959–62
Sir James Gray, FRS (1891–1975), 1962–5
Lord Platt, MD (1900–1978), 1965–8
Sir Alan Parkes, FRS (1900–1990), 1968–70
Mr. Peter R. Cox, 1970–2
Professor Cedric O. Carter, DM (1917–1984), 1972–6
Professor W. H. G. Armytage, 1977–82
Professor Bernard Benjamin, 1983–7
Professor Margaret B. Sutherland, 1987–9

Galton Institute 1989–

Professor Margaret B. Sutherland, 1989–93

Index

229